Recent Advances in Montmorillonite

Edited by Gustavo Morari Do Nascimento

Published in London, United Kingdom

Recent Advances in Montmorillonite
http://dx.doi.org/10.5772/intechopen.111054
Edited by Gustavo Morari Do Nascimento

Contributors
Abdesslem Ben Haj Amara, Amin Salhi, Asmae Charki, Ayoub Bayoussef, Bahareh Baheri, Daniil Vetyugov, Gustavo Morari do Nascimento, Hafsia Ben Rhaiem, Hassan Amhamdi, Hossain El Ouarghi, Iuliana Dogaru, Mahin Shahlari, Mihaela Stănciucu, Mohamed Amine Djebbi, Mohammed Mansori, Mourabit Fouad, M'hamed Ahari, Najlae Zaki, Nouhaila Hadoudi, Ramzi Chalghaf, Rihem Jemai, Saber Boubakri, Sonia Naamen, Sunggyu Lee, Tamara Matveeva

First published in London, United Kingdom, 2024 by IntechOpen
IntechOpen is the global imprint of INTECHOPEN LIMITED, registered in England and Wales, registration number: 11086078, 167-169 Great Portland Street, London, W1W 5PF, United Kingdom

British Library Cataloguing-in-Publication Data
A catalogue record for this book is available from the British Library

Additional hard and PDF copies can be obtained from orders@intechopen.com

Recent Advances in Montmorillonite
Edited by Gustavo Morari Do Nascimento
p. cm.
Print ISBN 978-0-85466-725-3
Online ISBN 978-0-85466-724-6
eBook (PDF) ISBN 978-0-85466-726-0

For EU product safety concerns:
IN TECH d.o.o., Prolaz Marije Krucifikse Kozulić 3, 51000 Rijeka, Croatia,
info@intechopen.com or visit our website at intechopen.com.

Meet the editor

Dr. Gustavo Morari do Nascimento has a position as an associate professor at the Federal University of ABC. He has a strong background and experience in many fields related to the characterization of advanced materials. Since his doctoral degree from the University of São Paulo (USP) followed by a postdoc at Massachusetts Institute of Technology (MIT), he has worked with carbon allotropes, conducting polymers and conducting polymer–clay nanocomposites. Nowadays, his research focus is on molecular characterization of chemically modified carbon nanostructured materials with molecular magnets and organic chalcogenide compounds by using FT-Raman, resonance Raman, surface-enhanced Raman spectroscopy (SERS), resonance Raman imaging, X-ray photoelectron spectroscopy (XPS), and X-ray absorption techniques at National Synchrotron Light Laboratory. Nowadays, he is also dedicated to studies in the history of chemistry, mainly on institutional, instrumental, and evolutionary concepts and concerns.

Contents

Preface

The synthesis and study of clays and their polymer–clay nanocomposites are one of the widest fields of research in materials science and technology. Characterization followed by application of many derived clay materials with different polymers and organic and biological molecules is one of the major features in clay science and technology. Nowadays, novel bidimensional materials based on clays and clay minerals warrant further applications from environmental to industrial concerns. As examples, the applications include tissue engineering, environmental remediation, biosensors, filtration, wound dressings, drug delivery, enzyme immobilization, geotechnical concerns, 3D printing, and so on. In this book, some chapters will bring new results concerning the state-of-the-art of synthesis, characterization, and applications. The main goal of this work is to contribute to the rationalization of some important results obtained in the clay science and technology area. The overall idea of the book is to provide the reader with a comprehensive and up-to-date evidence-based overview of the current literature in clay science and technology. This book discusses the main and emerging aspects of clay science and technology and is composed of an introductory section followed by two main parts: (1) Introduction, (2) "General Aspects of Clays and Polymer–Clay Nanocomposites", and (3) "Environmental and Industrial Applications of Clays and Polymer–Clay Nanocomposites".

The introductory chapter is focused on pillarized Montmorillonite clay as a possible host for new materials polymer-clay based materials.

Section 2: "General Aspects of Clays and Polymer–Clay Nanocomposites"

The second section of the book focuses on chapters dealing with fundamental properties and characterization of clays and their polymer-clay materials.

Chapter 2 brings a broad revision about the general properties of montmorillonite clays (such as structure, origin, chemical composition) and physical-chemical behavior (such as cation exchange, hydration and dehydration, absorption, surface capacity). These materials have a high potential to remediate the aqueous pollution from dyes, pharmaceutical derivatives, and many other applications.

Chapter 3 brings the characterization of polymer–clay nanocomposites mixed in polymeric blends. The main point is the preparation of the final material in a sequential mixture of the reagents. The organo-clay nanocomposites were characterized by instrumental techniques, such as TEM, XRD, TG, and rheological data, and their chemical and physical properties were determined. The enhancement of dispersion of clay particles in the polymeric blend improves the physical properties of the material and can be a prototype method for similar materials.

Section 3: "Environmental and Industrial applications of Clays and Polymer–Clay Nanocomposites"

The third section of the book focuses on the chapters dealing mainly with applications of clays and their related materials on environmental and industrial concerns.

Chapter 4 brings a synthesis and characterization of the CTAB-modified bentonite clay. The authors used several instrumental techniques in the characterization of the material. The modified clay shows improved properties in environmental remediation.

Chapter 5 brings a broad revision about the swelling properties of common clays (such as montmorillonite type) and the impact of their properties in the micro and macro scale range and their potential geotechnical hazard. The chapter is well elaborated and could be very useful for a broad audience. In fact, the methodological procedure for swelling analysis of clays has a potential broad application in the geochemical and correlated fields

Chapter 6 brings some important records related to the use of bentonite clays combined with organic polymer content to the processing of iron ores and pellets.

The knowledge regarding the real impact of materials derived from clays on environment and industrial concerns is still limited, despite the growth of the field in last decades. Nevertheless, I hope that this book will provide a small brick on this large investigation wall. Finally, I would like to thank all the authors for their contributions and the editorial manager for supporting the whole process.

Dr. Gustavo Morari do Nascimento
Professor,
Centre for Natural Sciences and Humanities,
Federal University of ABC,
Santo André, Brazil

Section 1

Introduction

Chapter 1

Introductory Chapter: Pillarized Montmorillonite Clay – A Possible Host for New Materials

Gustavo Morari do Nascimento

1. Introduction: pillarized montmorillonite clay - Synthesis and characterization

In the last three decades, many efforts have been made to the synthesis of conducting polymers with improved thermal, mechanical, and electrical properties. A promising way to acquire this purpose is by using clays as hosts of formed polymers or place to be used for the intercalation of monomers followed by polymerization [1–3]. Mainly smectite clays have been used when the confinement of the polymeric chains is desirable. The smectite clays are structurally formed from $MO_4(OH)_2$ groups of octahedral symmetrically bound to two MO_4 tetrahedral sheets producing layers designated T:O:T. The octahedral sites have ions such as aluminum, magnesium, and iron, while the tetrahedral centers accommodate silicon and aluminum. As a result, the layers have a negative charge with a parallel orientation, and the electric charge is neutralized by exchangeable hydrated positive ions in the interlayer space. This interlayer space has a lot of water molecules, both interlayer ions and water can be exchangeable with simultaneous change of the interlayer distance. These layered crystalline silicates have a great capacity for surface adsorption and catalytic activity in organic reactions, also including stabilization of radicals formed during polymerization (see **Figure 1**).

Montmorillonite (MMT) is the most used smectite clay, its negative layer charge arising mainly from the substitution of octahedral aluminum by magnesium (II) ions. Our group has been preparing and characterizing polymer-clay nanocomposites at least the last two decades [4–10]. The materials have been studied mainly by Raman spectroscopy and X-ray absorption techniques. Raman spectroscopy uses a laser as a source to probe the matter, the radiation inelastic scattered by samples is the Raman signal (stokes and anti-stokes components, see **Figure 2**). By the scattering process, the data about the vibrational modes can be extracted. Additionally, depending on the electronic structure of the sample, the Raman signal can be amplified by resonance and the vibrational states from excited states can be studied. Hence, by changing the laser energy, it is possible to investigate different structures of the intercalated monomer and polymer. Another powerful technique arises from a synchrotron radiation facility. In these machines, it is possible to tune the adequate photon energy values in relation to a specific element absorption edge. As a result, all possible atoms in the sample can be investigated separately in clay or its polymer-clay material by X-ray absorption techniques. An X-ray absorption spectrum (XAS) is a consequence of the core electron excitations to molecular unoccupied states (or extended states in the case

IntechOpen

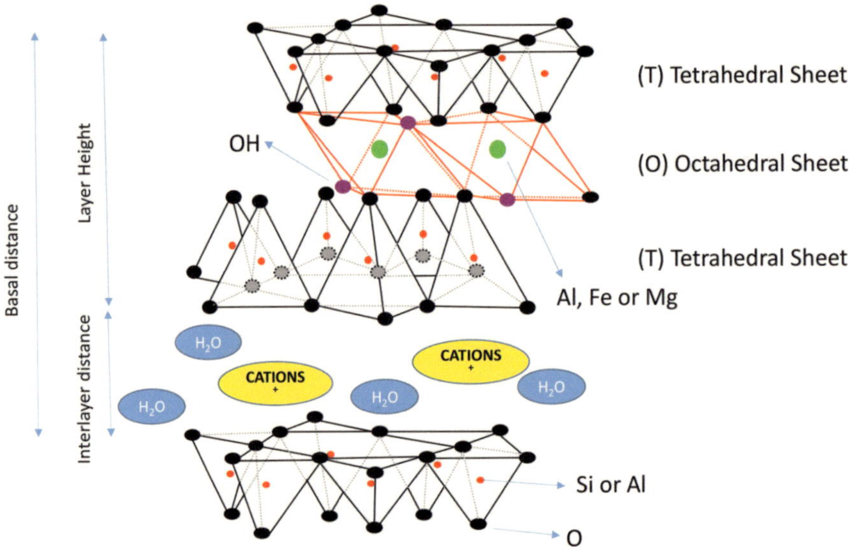

Figure 1.
Schematic representation of T:O:T structure of smectite clay group.

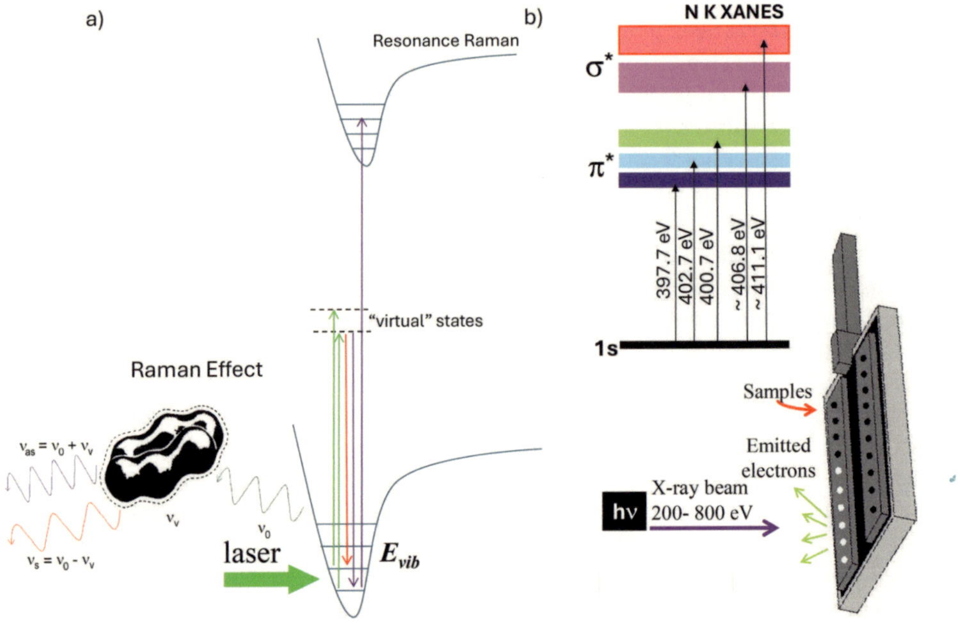

Figure 2.
(a) Schematic representation of ground and excited electronic states and their respective vibrational levels (the levels are out of scale). The arrows indicated transitions among levels. For Raman scattering, the laser line (ν_o) can be inelastically scattered by molecular vibrations and produce the Raman spectra (the scattered frequency is composed of stokes ν_s and anti-stokes ν_{as} components). If the laser line has energy like an electronic transition of material, the signal can be intensified by resonance process, this is the resonance Raman effect. (b) N K XANES experiments must be done in ultrahigh vacuum (the pressure inside the chamber is ca. 10^{-7} mbar). The measured signal is proportional to the absorption intensity [10, 11]. The arrangement used in our experiments was fully explained elsewhere [10, 11]. Schematic representation of the transitions from 1 s to π^* and σ^* states of a molecular material (the energy values are related to polyaniline in its emeraldine base form) is also displayed.

of solid samples). The absorption process at the N K shell (see **Figure 2**) can be visualized when the photon energy is transferred to a core electron to the atom with sudden changes in the absorption coefficient. The spectra show a lot of information about the electronic states of monomer/polymer and also the clay or other kinds of host.

Polymer-clay nanocomposites can exhibit better physical properties than their macroscopic composites. The possibility to make a real technological impact is the main reason for preparing new nanocomposites with clays. The polymer confinement into clay layers turns the polymeric chains more organized into nanoscale, as a consequence, this new arrangement at the molecular level creates specific supramolecular arrangements with an improvement of the polymer bulk properties. Several polymers have been synthesized into clay galleries, such as poly(ethylene oxide), poly-(styrene), and conducting polymers such as poly(aniline), poly(pyrrole), poly(*o*-methoxyaniline), and poly(2-ethynylpyridine) [12–16]. Although many efforts have been made to develop nanocomposites, nowadays is very difficult to predict and control their properties.

Pillarized clays are materials that have permanent porosity, obtained by the introduction of chemical compounds that function as molecular pillars between the clay lamellae, keeping them apart and giving rise to micropores. The chemical compounds that function as supports, or molecular pillars, between the clay lamellae are called pillaring agents (see **Figure 3**). The simple introduction of the pillaring agent by ion exchange gives rise to intercalated clays. In our study, the Keggin ion Al_{13}^{7+} was used to pillarize the MMT clay (see **Figure 3**). The study of pillarized clays began in the 1970s with the global oil crisis, due to the need to search for materials potentially applicable in oil cracking that had larger pores than those of the well-known zeolites [17, 18].

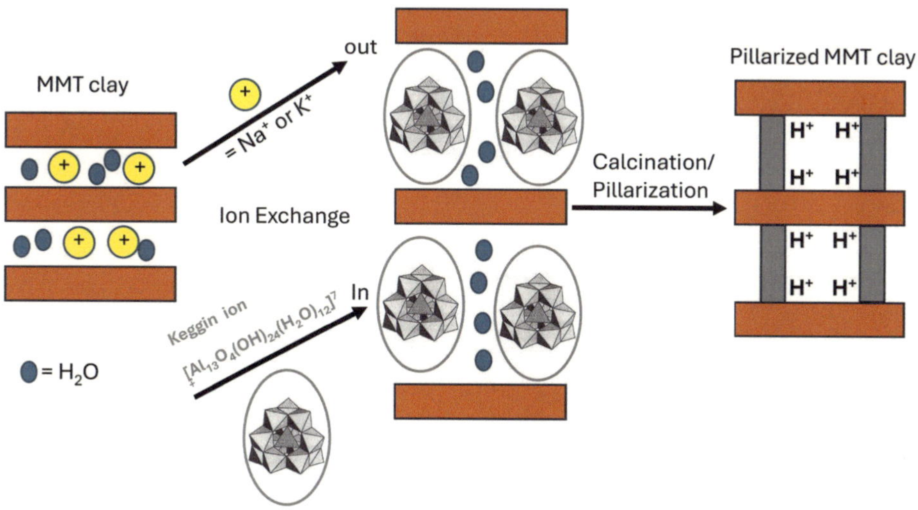

Figure 3.
Schematic representation of pillarization process of MMT clay. First step: ion exchange, sodium or potassium ions are replaced by Keggin ions Al_{13} polycations ($[Al_{13}O_4(OH)_{24}(H_2O)_{12}]^{7+}$ [4]. Typically, MMT-Al_{13}^{7+} solution can be prepared by a mixture of a volume of 100.0 mL of an aqueous solution containing the Al_{13}^{7+} cation (see Refs. [17–19] for preparation of solution containing Al_{13}^{7+} ions) with 140.0 mL of an MMT clay aqueous suspension (see Refs. [4, 5] for preparation of MMT suspension) under stirring and heated at 50°C. Afterward, the resulting suspension was kept under stirring and heating for another 6 h. Then, the suspension was filtered (0.22 μm Millipore filter), and the solid material was washed with deionized water and dried in a vacuum desiccator with silica gel. Second step: After intercalation, the modified clay is calcinated and there is the formation of stable pillars between the layers.

The objective of the pillarization process is to provide microporosity to the clay, creating materials containing pores with dimensions complementary to those of zeolites, that is, larger than 7 and smaller than 20 Å. If the pillaring agents are distributed homogeneously over the surface of the lamellae, a two-dimensional channel system will be created. It is also necessary that the clay lamella is rigid and does not bend; that the adsorption of pillaring agents on the external surface is negligible, and that all clay lamellae are pillarized [19].

The pillarization process can be characterized by two main techniques: X-ray diffraction (XRD) and Thermogravimetric (Tg) analysis. For instance, **Figure 4** shows the X-ray diffraction patterns of the Na^+-MMT clay before and after the pillarization procedure. It is clearly noted the shift of the d_{001} peak (see Table inside **Figure 4**) of MMT clay occurs after the pillaring method. This displacement corresponds to an increase in the interlayer space of 4.5 Å, confirming the success of the pillarization

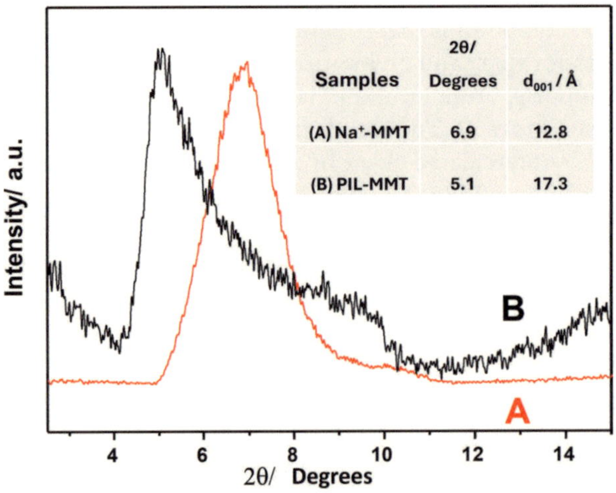

Figure 4.
XRD patterns of powdered samples of (a) Na + –MMT and (b) PIL-MMT. Inside the figure is placed a table with d_{001} values observed for both samples. XRD patterns of powdered samples were obtained on a Rigaku diffractometer model Miniflex using Cu Ka radiation (1.541 Å, 30 kV, 15 mA, step of 0.05°).

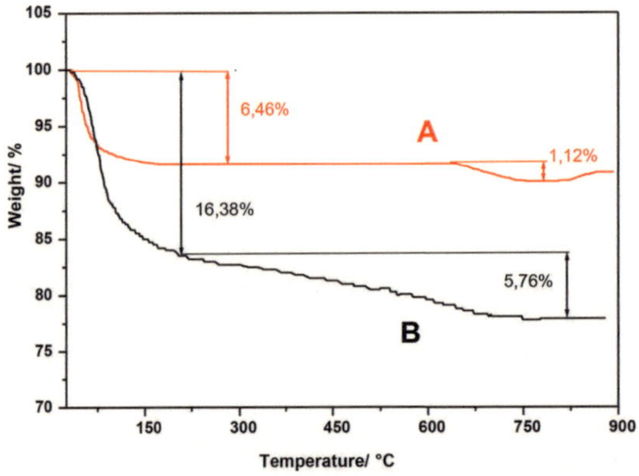

Figure 5.
TG curves of (a) Na^+-MMT and (b) PIL-MMT powdered samples. The data were acquired in a Shimadzu TGA-50 by using 10°C/min heating rate and synthetic air atmosphere.

process. Thermogravimetry analysis (Tg) can also be used to confirm the pillarization. For Na^+-MMT clay, there is a loss of water up to a temperature around 100°C, later stabilizing up to a temperature around 650°C, when dihydroxylations occur in its structure (see **Figure 5**). In pillarized clay, there is a loss of water up to a temperature of around 250°C, when successive dihydroxylations occur whereby the pillar is destroyed, proving pillarization in the clay (see **Figure 5**).

2. Polyaniline-pillarized montmorillonite clay: an example for new material

Nowadays, the preparation of polymer-clay nanocomposites or materials is an exemplary case of novel bidimensional materials based on clays and clay minerals that will warrant further applications from environmental to industrial concerns. Here,

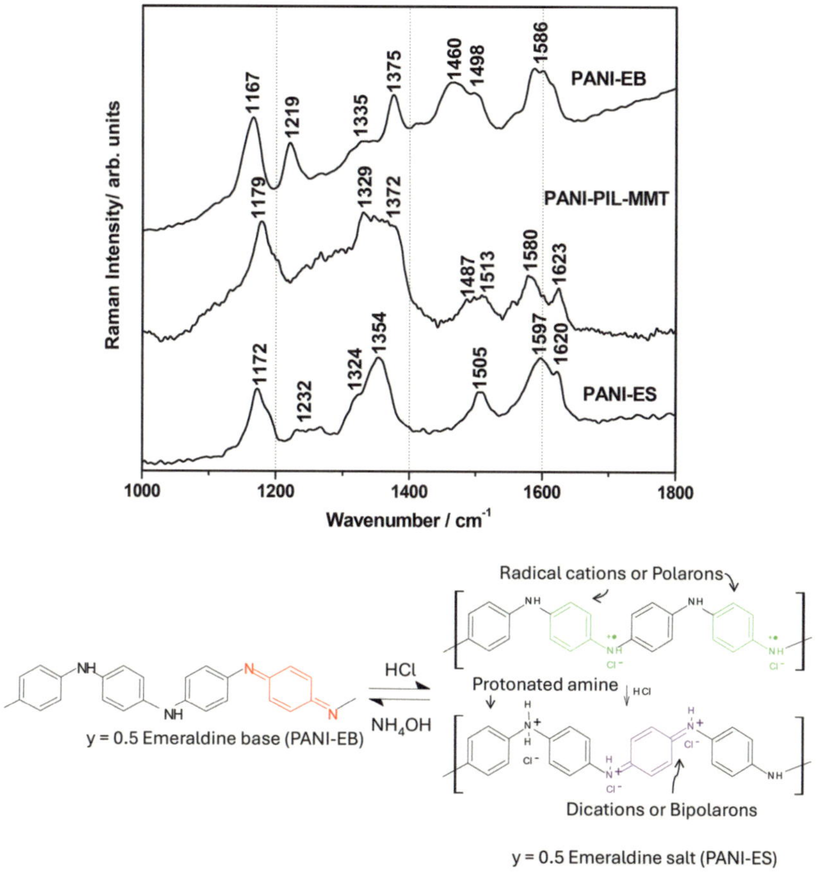

Figure 6.
FT-Raman spectra of powdered samples of PANI-ES, PANI-PIL-MMT, and PANI-EB. PANI has different oxidation and protonation states; the emeraldine base state (PANI-EB) is in which the number of reduced or benzenoid units is equal to oxidized or quinoid units. The conducting form of PANI as the emeraldine salt (PANI-ES) has benzenoid, radical cation, and dications segments (see Scheme) [20]. FT-Raman spectra (1064.0 nm laser line) were acquired from powdered samples in a Bruker RFS 100/S FT-Raman spectrometer at room temperature (near 20°C).

we will show an example that has been studied in our lab and can be replicated for many other cases.

Aniline was added to the PIL-MMT aqueous and then a suspension of An^{+}-PIL-MMT was polymerized with potassium persulfate. The solid material (PANI-PIL-MMT) was then characterized by spectroscopic techniques. The FT-Raman technique has been used in our group for systems where the fluorescence is very high. For some clays and materials derived from clays, there is the presence of high intrinsic fluorescence, this is the case for PIL-MMT samples and derived materials. FT-Raman uses exciting radiation close to the infrared region (1064.0 nm) and the problem of fluorescence is bypassed for a major part of cases.

For instance, **Figure 6** shows that there is a shift in the bands of the polymer obtained in the pillared clay when compared to the Raman spectrum of polyaniline emeraldine-salt (PANI-ES)—both obtained with exciting radiation in 1064 nm. It is noted that the whole spectrum pattern of PANI-PIL-MMT is like the PANI-ES, the highly doped PANI form, spectrum. Thus, the bands at 1179, 1329, 1372, 1515, and 1623 cm^{-1} can be associated to the similar bands from PANI-ES at 1172, 1324, 1354, 1505, and 1620 cm^{-1}. For PANI-ES, bands from 1324 to 1354 cm^{-1} can be assigned to νC–N modes of polarons, and these νC–N bands shift to 1326–1375 cm^{-1} when the PANI-ES is submitted to mild deprotonation [21], as a consequence of the formation of small conjugated polarons.

However, two bands at 1487 and 1580 cm^{-1} in the Raman spectrum of PANI-PIL-MMT can be related to the PANI-EB sample (deprotonated form of PANI) at 1460 and 1586 cm^{-1}. These bands are attributed to νC=N and νC=C of quinoid groups [21]. Hence, the PANI was obtained into PIL-MMT probably in a form intermediated to the PANI-ES and PANI-EB forms. It is important to stress that the values of bands of inserted PANI into PIL-MMT can also be affected by the interaction between PANI chains and PIL-MMT walls [22–24]. In the UV-vis-NIR (not shown), it is not noted PANI bands, however, this behavior is due to two major factors: low stability of PANI-PIL-MMT aqueous suspension and low formation of PANI into PIL-MMT (despite visual color change of sample before and after polymerization).

3. Conclusion and future remarks

In this brief chapter, we give an introductory overview of the use of pillarized MMT clay as a host for the synthesis of polymer-clay nanocomposites. Considering this example and many other cases related to our group, there are two central questions that are possible to address: (i) the molecular structure of the polymer is drastically changed inside the interlayer cavity of clay and (ii) by using the appropriate synthetic or heating route is possible to change the molecular structure of the confined polymer. X-ray diffraction and thermogravimetry can be used in combination to confirm the successful pillarization of the montmorillonite clay. Nowadays, we are changing the experimental conditions to maximize the PANI formation, but many other systems can be prepared following this suggestion. We really think that the polymer-clay area has much more to be investigated. In the last two decades, our group has deeply studied PANI-MMT materials and derivates. These studies have revealed the importance of the use of advanced and complementary spectroscopic techniques to give a more realistic view of the molecular structure formed into the clay layers. The chains formed inside the clay layers can present different segments and molecular arrangements compared to the free polymer. The new Raman instruments can give better Raman imaging of the samples, and open the possibility to study inhomogeneity, chemical modifications and many other aspects of the polymer-clay materials. In addition, new Synchrotron light sources will permit to study changes in these complex materials by *in situ* measurements, resulting in spectral and imaging data. These new data are crucial for better application design and more vast use of materials derived from polymer-clays.

Author details

Gustavo Morari do Nascimento
Federal University of ABC, CCNH, Santo André, Brazil

*Address all correspondence to: gustavo.morari@ufabc.edu.br; morari@yahoo.com

References

[1] Yariv S. Introduction to organo-clay complexes and interactions. In: Yariv S, Cross H, editors. Organo-Clay Complexes and Interactions. New York: Marcel Dekker, Inc.; 2002

[2] Pandey JK, Reddy KR, Kumar AP, Singh RP. An overview on the degradability of polymer nanocomposites. Polymer Degradation and Stability. 2005;**88**:234

[3] Hussain F, Hojjati M, Okamoto M, Gorga RE. Review article: Polymer-matrix nanocomposites, processing, manufacturing, and application: An overview. Journal of Composite Materials. 2006;**40**(17):1511

[4] Do Nascimento GM, Constantino VRL, Temperini MLA. Spectroscopic characterization of a new type of conducting polymer-clay nanocomposite. Macromolecules. 2002;**35**:7535

[5] Do Nascimento GM, Landers R, ConstantinoVRL,TemperiniMLA.Aniline polymerization into montmorillonite clay: A spectroscopic investigation of the intercalated conducting polymer. Macromolecules. 2004;**37**:9373

[6] Do Nascimento GM, Constantino VRL, Temperini MLA. Spectroscopic characterization of doped poly(benzidine) and its nanocomposite with cationic clay. The Journal of Physical Chemistry B. 2004;**108**:5564

[7] Do Nascimento GM, Landers R, Constantino VRL, Temperini MLA. Spectroscopic characterization of polyaniline formed in the presence of montmorillonite clay. Polymer. 2006;**47**:6131

[8] Do Nascimento GM. Structure of clays and polymer-clay composites studied by X-ray absorption spectroscopies. In: Do Nascimento GM, editor. Clays, Clay Minerals and Ceramic Materials Based on Clay Minerals. 1st ed. London: InTechOpen; 2016

[9] Do Nascimento GM. Introductory chapter: Polymers and clays - A fruitful combination. In: Do Nascimento GM, editor. Clay and Clay Minerals. 1st ed. London: InTechOpen; 2021

[10] Do Nascimento GM. Two spectroscopies as Main source for investigation of polymer-clay materials. In: Do Nascimento GM, editor. Clay Science and Technology. 1st ed. London: InTechOpen; 2021

[11] Do Nascimento GM. X-ray absorption spectroscopy of nanostructured polyanilines. Chemical Papers. 2013;**67**:933

[12] Wu Q, Xue Z, Qi Z, Hung F. Synthesis and characterization of PAn/clay nanocomposite with extended chain conformation of polyaniline. Polymer. 2000;**41**:2029

[13] Biswas M, Ray SS. Water-dispersible nanocomposites of polyaniline and montmorillonite. Journal of Applied Polymer Science. 2000;**77**:2948

[14] LeBaron PC, Wang Z, Pinnavaia TJ. Polymer-layered silicate nanocomposites: An overview. Applied Clay Science. 1999;**15**:11

[15] Meneghetti P, Qutubuddin S. Synthesis, thermal properties and applications of polymer-clay nanocomposites. Thermochimica Acta. 2006;**442**:74

[16] Armentano I, Dottori M, Fortunati E, Mattioli S, Kenny JM. Biodegradable polymer matrix nanocomposites for tissue engineering: A review. Polymer Degradation and Stability. 2010;**95**(11):2126

[17] Heylen I, Vanhoof C, Vansant EF. Preadsorption of organic compounds on iron oxide-pillared clays. Microporous Materials. 1995;**5**:53-60

[18] Luna FJ, Schuchardt U. An introduction to pillared clays. Química Nova. 1999;**22**(1):104

[19] Liu G-H, Ko A-N, Chang Y-C. Synthesis and properties of pillared montmorillonite formed by intercalation of transition metal macrocyclic complexes. Microporous Materials. 1995;**5**(1-2):61-67

[20] MacDiarmid AG. 'Synthetic metals': A novel role for organic polymers (Nobel lecture). Angewandte Chemie, International Edition. 2001;**40**:2581

[21] Do Nascimento GM, Temperini MLA. Studies on the resonance Raman spectra of polyaniline obtained with near-IR excitation. Journal of Raman Specroscopy. 2008;**39**:772-778

[22] Do Nascimento GM, Pradie NA. Deprotonation, Raman dispersion and thermal behavior of polyaniline-montmorillonite nanocomposites. Synthetic Metals. 2016;**217**:109

[23] Do Nascimento GM, Temperini MLA. Structure of polyaniline formed in different inorganic porous materials: A spectroscopic study. European Polymer Journal. 2008;**44**:3501

[24] Do Nascimento GM, Souza MA. Spectroscopy of nanostructured conducting polymers. In: Eftekhari A, editor. Nanostructured Conducting Polymers. Londres: Wiley and Sons; 2010. pp. 341-375

Section 2

General Aspects of Clays and Polymer–Clay Nanocomposites

Chapter 2

Montmorillonite: Properties, Characteristics, and Its Harnessing in Environmental Applications

Rihem Jemai, Ramzi Chalghaf, Saber Boubakri, Mohamed Amine Djebbi, Sonia Naamen, Hafsia Ben Rhaiem and Abdesslem Ben Haj Amara

Abstract

Montmorillonite (MMT) is a 2:1 dioctahedral clay mineral. Due to all the environmental uses, it is an adaptable clay mineral that garnered much attention. In this chapter, we will highlight the distinctive properties and special qualities of MMT and its crucial function in environmental contexts. In this chapter, we will also explore its remarkable adsorption powers to remove contaminants from water and its contribution to erosion reduction and soil stabilization. This chapter reveals through an in-depth examination how MMT promotes eco-friendly practices and offers effective solutions for reducing pollution and preserving ecological balance. Then, we merged the wide range of characteristics of MMT with the vast array of properties of starch biopolymer to prepare a nanocomposite with a high adsorption capacity. Thus, we found a large adsorption capacity of 341.9 mg/g with a removal rate of 97.7%. Furthermore, we will explore its efficacy in removing ibuprofen drug by modifying it with CTAB/ZnO nps. Additionally, its potential application in the electrochemical detection of NiO_2^- will be investigated through the incorporation of TiO_2 and ZnO nps on MMT.

Keywords: montmorillonite, properties, adsorption, environment, water purification, contaminants

1. Introduction

The MMT is a clay mineral composed of two silica tetrahedral sheets and one alumina octahedral sheet that make up this 2:1 layer phyllosilicate. It belongs to the dioctahedral smectite group. Because of its large specific area, high cationic exchange capacity, and indeed its high adsorption capacity and high porosity, MMT is thought to be as a suitable candidate for many applications. For instance, in drug delivery systems, MMT is employed as a pharmaceutical excipient. Its strong adsorption ability contributes to the improvement of drug entrapment and sustained release. In many formulations, MMT usually maintains medication release by firmly adhering to the drug molecule. Furthermore, MMT increases the rate of hydrophobic drug dissolution and bioavailability [1]. It is found in various environmental applications, including soil remediation and water purification, and is a constituent of environmentally friendly materials [2]. Thus, it is used to adsorb toxic compounds like heavy metals such as $Hg^{2+}, Cr^{3+}, Pb^{2+}, Cu^{2+}, Zn^{2+}, Ba^{2+}, Ni^{2+}, Mn^{2+}, Cd^{2+}$ and Ag^{+} [3], synthetic dye as methylene blue [4] and pharmaceutic compounds [5]. However, when compared to other adsorbent materials, MMT might have a comparatively low adsorption capability when used alone. This capability may differ based on several variables, such as specific surface area and capacity of cation exchange. MMT's specific surface area, the ratio of its surface area to mass, has a major effect on how well it adsorbs different compounds. Increasing the specific surface area of MMT can significantly increase its adsorption capacity [6]. Furthermore, the many kinds of adsorption sites that are found on the surface of MMT are crucial in determining how well it can adsorb substances. Interestingly, certain adsorption sites in MMT draw different kinds of molecules, which increases the material's capacity to adsorb organic and inorganic compounds [6]. Additionally, the adsorption capacity of MMT is influenced by its chemical composition. It is possible to significantly increase MMT's adsorption capacity through chemical changes, allowing for a wider variety of adsorbed compounds, as demonstrated in the context of chemical alteration [7].

It underwent an acidic treatment to enhance its specific area and then its adsorption capacity. For example, as indicated by Ravichandran and Sivasankar, MMT had an external specific surface area of 19.0 m^2/g [6, 7]. However, after the HCl (0.7 M) treatment, its value dramatically increased to 188.3 m^2/g [7, 8]. The treated MMT was then used to adsorb Pb(II), Cd(II), Cu(II), Co(II), and Ni (II) from water [8]. Adding nanoparticles to MMT can also enhance specific surface area by generating more adsorption sites and changing the structure of MMT [9]. It can be metallic nanoparticles such as zinc oxide nanoparticles [10], TiO_2 nanoparticles [11], etc. The modification of MMT by organic products and polymers is another commonly used method for organic products, including solvents such as alcohols, ethers, and ethanol [12]. It can also be modified with plastic polymers such as polyethylene [13] and polystyrene [14] or biopolymers such as chitosan [15], starch [16], silane organic chains, alkyls, amines [16].

In this chapter, we will explore the diverse properties of MMT, rendering it indispensable across various fields. Emphasis will be placed on its intrinsic qualities, showcasing how this naturally occurring mineral is not only supportive of environmentally friendly practices but also presents viable solutions for enhancing water quality, reducing pollution, and preserving nature. A detailed examination will be conducted, focusing on the adsorption of methyl orange dye using MMT modified by octadecylammonium and starch biopolymer [17], removal of ibuprofen by MMT-CTAB/ZnO [18] np with photocatalysis methods, and electrochemical detection of NiO_2^- by MMT-TiO_2/ZnO [19], delving into the mechanisms and outcomes involved.

2. Structure of MMT clay mineral

The MMT mineral is a 2:1 phyllosilicate. It is composed of two SiO_4 tetrahedral sheets sandwiching an octahedral sheet [20]. This mineral structure exhibits a dioctahedral phyllosilicate composition wherein, among every three octahedral cavities, two are occupied by a trivalent cation (Al^{3+}) [21]. Two planes of oxygen and hydroxide ions organized in a compact assembly formed the octahedral sheet, delineating octahedral cavities—two-dimensional networks of tetrahedra form adjacent to this

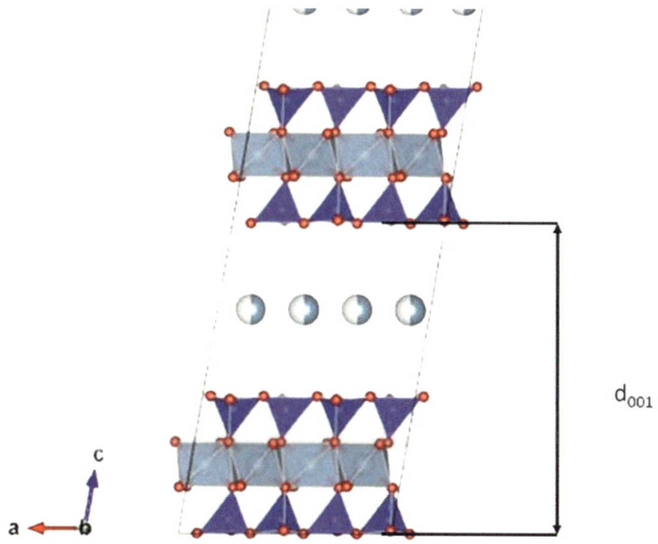

Figure 1.
MMT crystal structure by VESTA software projected along (010).

center sheet. The assembly of octahedral and tetrahedral sheets forms a layer. The space between two layers is called interfoliar space. The thickness of the layer with interfoliar space forms the basal spacing d_{001} (**Figure 1**). MMT can be identified by a doubly periodic repeat of a planar lattice that is centered and has parameters 'a' and 'b', where $b = a\sqrt{3}$. The structure has monoclinic symmetry ($\alpha = \gamma = 90°$, $\beta \neq 90°$) due to the orientation of the two tetrahedral sheets and a space group C2/m [21]. Moreover, MMT exhibits stacking faults, including faults due to translations and random rotations [21]. Generally, all types of phyllosilicates have the same lattice parameters (a, $b = a\sqrt{3}$) and variable c parameter. This is why we always consider d_{001} when referring to basal spacing.

Dioctahedral smectite is formed of an octahedral cavity layer around two SiO_4 tetrahedral layers [22]. By stacking oxygen planes, tetrahedral or octahedral cavities were created, resulting in tetrahedral or octahedral layers, respectively. The stability of the layer is maintained by the presence of cations within these cavities. MMT is a 2:1 dioctahedral phyllosilicate in which a trivalent metal ion (Al^{3+}) occupies two of the three octahedral cavities. In the tetrahedral layer ($Si^{4+} \rightarrow Al^{3+}$) or the octahedral layer ($Al^{3+} \rightarrow Mg^{2+}$, $Mg^{2+} \rightarrow Li^{+}$), isomorphic substitutions can occur [22]. The resulting layers are then negatively charged, and to maintain sheet's electrical neutrality, compensating cations are placed in the distance between the layers (d_{001}).

The clay fraction is composed of three structural units (**Figure 2**): the layer (7–25 Å), the particle (200–1500 Å), and the aggregate (1.5–16 μm) [24]. This structuring determines two types of porosity: inter-aggregate porosity and intra-aggregate porosity.

MMT is studied and characterized at various scales using diverse methods. At the nanometric scale, powder X-ray diffraction (PXRD) and small angle X-ray scattering (SAXS) are employed to determine cation position, average of water layer, and stacking modes, which can characterize MMT structure when modified with inorganic or organic compounds and to identify structural changes along c parameter [21].

Fourier transform infrared spectroscopy (FTIR) (500–4000 cm^{-1}) is used to determine the exchange capacity ratio and the dioctahedral type of MMT. Inductively

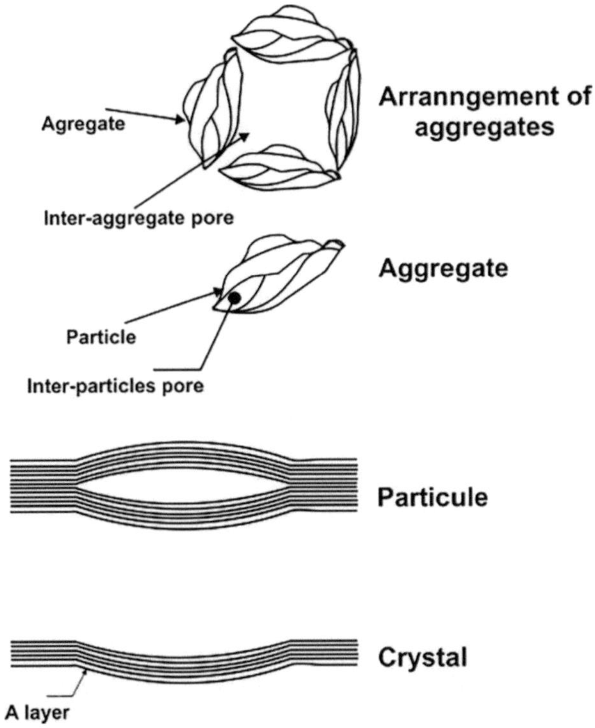

Figure 2.
The various types of pore spaces in MMT at different scales [23].

coupled plasma (ICP) is utilized to determine the chemical elements and the amount of oxides in MMT. Electronic microscopy is employed to identify and monitor the morphology of both raw and treated MMT. In fact, it enables the observation of MMT at the particle scale, allowing us to visualize the pores and layers [23].

3. Types of MMT

Multiple types of MMT were distinguished among them via numerous factors:

3.1 Chemical composition

The general formula for MMT is provided as follows [25]:

$$(Si_4)\left(Al_{2-y}\,Mg_y\right)O_{10}\,(OH)_2\,,\,{}_yM^+.nH_2O$$

The chemical composition of MMT varies slightly depending on where it comes from. Its characteristics may change depending on the elements or cations present and their concentration. Thus, when metal ions bind to the adsorption sites of MMT, they have the potential to replace some of the Al, Ca, or Mg cations within the MMT structure. This can lead to the formation of new minerals with different chemical compositions, subsequently influencing the color of MMT [26].

Place	Chemical formula
Wyoming/USA [23]	$(Ca^{2+}{}_{0.07}K^{+}{}_{0.01}Na^{+}{}_{0.20})\,[Al^{3+}{}_{1.53}Fe^{3+}{}_{0.18}Mg^{2+}{}_{0.26}Ti^{4+}{}_{0.01}]$ $[Si^{4+}{}_{3.96}Al^{3+}{}_{0.04}]\text{-}O_{10}(OH_2)$
France [26]	$(Ca_{0.14}Na_{0.02})_{\Sigma=0.16}\ (Al_{1.66}Mg_{0.36}Fe_{0.04})_{\Sigma=2.08}\ (Si_{3.90}Al_{0.10})_{\Sigma=4.00}O_{10}$ $(OH)_2\ 1.02H_2O$
Choushan, Zhejiang Province (China) [27]	$(Ca_{0.10}Mg_{0.08})(Al_{1.51}Fe^{3+}{}_{0.14}Mg_{0.35})(Si_4)O_{10}(OH)_2{\cdot}nH_2O$
Gafsa (Tunisia) [24]	$(Si_4)(Al_{1.1}\ Mg_{0.4}\ Fe_{0.5})\ O_{10}\ (OH)_2\ M^{+}$ (M is a compensating cation)
Grec-Ca [23]	$(Si_{3.923}\ Al_{0.077})(Al_{1.459\ 26}Ti^{4+}{}_{0.018}\ Fe^{3+}{}_{0.039}Fe2^{+}{}_{0.045}Mg_{0.382})\ O_{10}\ (OH)_2$ $(Ca^{2+}{}_{0.18}\ Na^{+}{}_{0.03})$

Table 1.
MMT types according to geological formation.

3.2 Geological origin

The physical and chemical properties of MMT might vary depending on the geological formations or mineral deposits. A naturally occurring mineral, montmorillonite can be found all over the world. It was first discovered in France in 1847 at Montmorillon [25], hence its name. It is difficult to get a precise estimate of the total amount of MMT in the world. The amount of MMT in the earth is determined by a number of factors, such as the concentration of the mineral in the soil, the amount of soil that has been studied and examined, and the amount of MMT that is artificially produced. Furthermore, because MMT is a mineral that readily dissolves into soil, accurate assessment is frequently difficult. All these conditions lead to slight substitutions in the tetrahedral sheet charge, with the charge varying e of 0 and 0.1. Based on their chemical formulae, the regions where MMT is found are listed in **Table 1**.

3.3 Purity and impurities

Impurities may affect texture, color, and intended uses of MMT clay minerals [27]. Such impurities like traces of heavy metals or other minerals like quartz and calcite, which are identified with XRD, could affect the MMT utilization.

4. Purification of MMT clay mineral

The purification process is very important to determine the structural formula before using the MMT clay mineral.

Generally, there are two stages for MMT purification: treatment with hydrochloric acid (HCl) and the use of NaCl (1 M). These protocols are used to ensure the removal of impurities completely.

The HCl titration method was used to remove $CaCO_3$ from clay minerals, but the pH must not be lower than 5 as this could damage the structure of the MMT clay mineral. On the other hand, the NaCl purification method allows maximum dispersion of the mineral and removal of all impurities greater than 2 μm, as well as cationic exchange. The purification protocol involved dispersing 1 g of MMT in 10 ml of a 1 M NaCl solution. The mixture was stirred using a magnetic stirrer for 24 h and subsequently centrifuged at 4000 rpm for 10 minutes. The cationic exchange reaction is represented as follows:

$$MMT - M^{+} + NaCl \rightarrow MMT - Na^{+} + MCl \quad (1)$$

The recovered solid was then washed many times with distilled water to ensure the entire removal of chloride ions. Finally, it was dried at 60° C at 12 h.

5. Cationic exchange capacity

All or part of the compensating cations can be exchanged with those of a saline solution in contact with the clay mineral. This process, known as cation exchange capacity (CEC), can repetitively occur with various cations in the solution. In order to understand the characteristics of these minerals, particularly their hydration and dispersion properties, introducing the concept of cation exchange capacity becomes necessary. Thus, the unique feature of each 2:1 phyllosilicate lies in its cation exchange capacity, denoting the number of monovalent cations capable of substituting the compensating cations in a silicate to balance the electrical charge of 100 g of calcined mineral [23]. For soils (e.g.), the CEC represents the soil's reservoir of fertility.

The CEC measurements are conducted using various protocols such as the titration method [28–30], employing ethylene diamine tetra-acetic acid (EDTA) on MMT-Na converted to its calcium form and subsequently reverted to its sodium form to measure the calcium ions that were liberated [30]. Additionally, the Hofman-Klemen test is employed to measure the CEC involving the loss of the entire cation exchange capacity of MMT saturated with a cation possessing a small ionic radius, such as Li (0.63 Å), after heating to 200°C due to the migration of Li cations to the vacant octahedral site [20, 31]. The CEC of MMT ranges between 80 and 150 meq/100 g [32].

6. Hydration and dehydration

6.1 Hydration

Investigating the hydration process in MMT is crucial to evaluate its ability to encapsulate pollutant molecules in its interlayer space. In fact, the hydration property is defined as the ability of MMT to swell in the presence of water.

The distribution of water in the MMT occurs either at the particle scale defining internal water that leads to the penetration of water into the layers of MMT (interlayer

MMT type	Basal distance d_{001} (Å)	Ion radius (Å)	Number of water layer	References
MMT-Na	≃11.6–12.9	1.12	1	[9, 34]
MMT-Cs	12.35	1.17	1	[35, 36]
MMT-Cu	12.36	0.71	1	[36]
MMT-Li	14.62	0.63	2	[37]
MMT-Ni	15.08	0.72	2	[36]
MMT-Co	14.88	0.74	2	[36]

Table 2.
Average water layer in various types of MMT.

hydration) or in the micropores (interparticle dispersion), or between particles and aggregates, referred to as external water [33]. The interlayer hydration varied according to several factors such as CEC, ionic radius of interlayer cations, pH, pressure, relative humidity (RH), and pH. The average number of water layer in MMT can be more than three layers under such conditions of RH, etc., leading to a change in basal distance [32] estimated through the X-ray diffraction method. In **Table 2**, some examples of the number of water layers in MMT are provided based on cation exchange and its ionic radius [32]. Meanwhile, the external water is estimated to have an average of 4.5 g in Wyoming-Ca [38].

6.2 Dehydration

MMT can undergo a dehydration process under specific conditions involving the removal of water molecules from the MMT structure through heating. The dehydration process was thoroughly examined using thermogravimetric analysis, revealing two distinct stages. The initial stage, occurring between 100°C and 140°C, is associated with the extraction of pore water (external water) due to the very low interaction energy between water molecules and the MMT structure. The second stage, observed in the temperature range of 500–600°C, corresponds to the dehydration of internal water, attributed to the high interaction energy between the water layer and cations [33]. Water distribution is generally conducted through the desiccation/humectation method. In a cycle of desiccation/humectation of MMT, the gain and loss of internal water molecules are quasi-reversible under certain conditions. However, it can be irreversible with smaller interlayer cations (such as lithium and magnesium) [37, 39]. The water content curve in relation to pressure exhibits hysteresis [37].

The concept of hydration and dehydration is necessary to understand the behavior of MMT and its application in various fields, such as water purification and soil remediation.

7. Adsorption capacity

Adsorption capacity of MMT depends on many factors such as surface area, swelling degree, capacity of cationic exchange, and surface charge. But sometimes, it must be modified to elevate adsorption properties. It can be modified with a cation by the addition of an ammonium or phosphonium-based surfactant to MMT, polymer chains, alkyl chains, amines [40, 41], grafting an organosilane [42], etc. The modification of MMT with an organic compound results in an increase in surface area and facilitates the intercalation of various compounds, including nanoparticles, polymers, catalysis, etc. This modification enhances the potential applications of MMT in environmental contexts **Table 3**. Adsorption capacity is closely related to the surface area when estimating adsorption sites in MMT.

8. Surface area

Various methods have been employed to estimate and determine the surface area and porosity of MMT clay minerals. The absorption of methylene blue (MB) dye is a commonly utilized procedure [4, 49]. Surface area measurement is also conducted using the Brunauer-Emmett-Teller method, involving the adsorption/desorption of nitrogen at 77 K [50]. Furthermore, investigations into the adsorption of ethylene glycol are carried out. The total surface area value is estimated at 700–800 m^2/g [51].

Compound	Method	Contaminant	Adsorption capacity	Ref.
MMT/chitosan nanocomposite	Adsorption	Methyl orange	123.46 mg/g	[43]
Kappa-carrageenan (Car)/ sodium alginate (Alg) biopolymers/Na-MMT	Adsorption	Crystal violet	88.8 mg/g	[44]
MMT/bean starch nanocomposite	Adsorption	Heavy metals: Co, Ni	Co: (94%) Ni: 78%	[45]
MMT/TiO_2 nanoparticle	Photocatalyse	Rhodamine B		[46]
Phenanthroline/MMT composite	Electrochemical detection	Pb		[47]
montmorillonite-ZnO	Electrochemical detection	Diltiazem hydrochloride drug		[48]

Table 3.
Various applications of MMT in water remediation.

9. Main characterization techniques of MMT clay mineral

Many techniques are used to identify and quantify MMT clay minerals. First, the use of powder X-ray diffraction (PXRD) is often considered the primary technique in an optimal strategy for MMT clay mineral identification. It relies on the application of X-rays to the powdered MMT clay mineral sample [25], following Bragg's law:

$$2dsin\theta = n\lambda \tag{2}$$

where:
λ : X-ray wavelength.
n: diffraction order
d : interplanar spacing
θ: angle of X-ray incidence.

This method allows the identification of the clay mineral structure because the dimension of X-ray wavelength is on the order of atom dimensions (angustrum). It also permits the determination of the position of exchanged cations in the MMT structure and characterizes the stacking modes [25, 38]. Second, the identification of MMT clay minerals is followed by the utilization of electron microscopy techniques such as scanning electron microscopy (SEM) and transmission electron microscopy. These methods enable the visualization of MMT clay mineral morphology and tracking changes that occur after any modifications. They also allow the determination of its chemical composition with energy dispersion X-ray fluorescence (EDXRF), its lattice imaging with high-resolution transmission electron microscopy (HRTEM) in TEM, and its structure by selected area electron diffraction (SAED) in TEM. Additionally, Fourier transform infrared spectroscopy (FTIR) is commonly used, and it is based on the absorption of IR radiation by the MMT sample. This technique involves the vibrations of atoms that constitute the clay mineral structure. By analyzing the positions of absorption bands, it is possible to detect functional groups and identify MMT clay minerals. With FTIR, it is easy to identify the dioctahedral type of

MMT. It can indicate exchanged cations and efficiently determines organic intercalated molecules [25]. There are many other complementary techniques such as X-ray photoelectron spectroscopy (XPS), X-ray Absorption Spectroscopy, etc.

10. Montmorillonite modified with octadecylammonium/starch composite (PSA/OMMT) for effective removal of methyl orange

The use of MMT clays to reinforce starchy matrices for packaging and wastewater treatment has been investigated in earlier research [52, 53]. Adding MMT modified with octadecylammonium cation to the porous starch matrix (PSA/OMMT) improves its capacity to remove dyes from aqueous solutions. This improvement is brought about by an increase in the specific surface area, which is made possible by better interactions between the organoclay and polysaccharide. These interactions increase the surface functionality of the composite by resulting from the exfoliation of the organoclay and possible hydrogen bond connection of biopolymer chains [54, 55]. **Figure 3** provides a graphical illustration of PSA/OMMT. Prior research has investigated the use of MMT clays to strengthen starchy matrices for various purposes.

The adsorption capacity of the produced PSA/OMMT composite was evaluated using the anionic dye methyl orange (MO).

The adsorption experiments were conducted by studying kinetic adsorption, isotherm adsorption, and thermodynamic adsorption, but first, many parameters were optimized, which are pH effect, time effect, and adsorption dosage effect.

The removal percentage and adsorption capacity experienced a notable increase at pH 9 with 98.5% and 189.7 mg/g [17], respectively. Consequently, pH 9 was

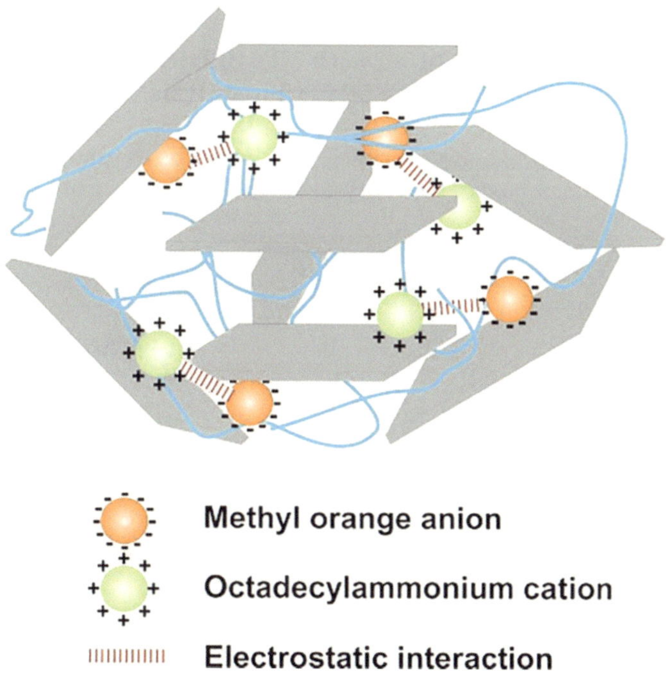

Figure 3.
Graphical representation of PSA/OMMT in MO adsorption.

Adsorbent compound	Adsorption capacity (mg/g)	References
Zn/Al layered double hydroxide	276.55	[56]
Biochar from grape seeds	111.11	[57]
Mesoporous carbon CMK-3	294.1	[58]
PSA/OMMT (1:10)	344.7	[17]

Table 4.
Adsorption of MO with various adsorbents mentioned in the literature.

determined as the most effective pH for subsequent experiments. Next, the investigation delved into the reaction time, spanning from 5 to 90 minutes, revealing a swift adsorption phase within 30 minutes with a removal rate of 99.2% and an adsorption capacity of 198.5 mg/g. Furthermore, the study of adsorbent dosage on MO adsorption showed a maximal removal efficiency of 197.7 mg/g (with 20 mg of adsorbent). At those optimized, the analysis of kinetic adsorption results indicated that the PSA/OMMT composite's adsorption of MO fitted well with the pseudo-second-order model. Then, the investigation of the adsorption isotherms results demonstrated a high fitting to the experimental data compared to the Freundlich model and a predicted adsorption capacity close to the experimental value (344.7 mg/g) [17]. This result suggests a preference for a uniform monolayer adsorption behavior for MO dye molecules. Finally, the study of the thermodynamic effect shows that the adsorption of MO on the PSA/MMT is an exothermic nature. Consequently, PSA/OMMT demonstrates significant potential as an efficient and low cost adsorbent for the treatment of methyl-orange-containing wastewater [17]. **Table 4** presents a comparative study of these works and other adsorbents for the adsorption of methyl orange.

11. The efficacy of MMT modified with CTAB/ZnO nanocomposite for the removal of methylene blue dye by photocatalysis

The incorporation of nanoparticles into MMT broadens its range of applications, especially in fields like photocatalysis and electrochemical detection. In this study,

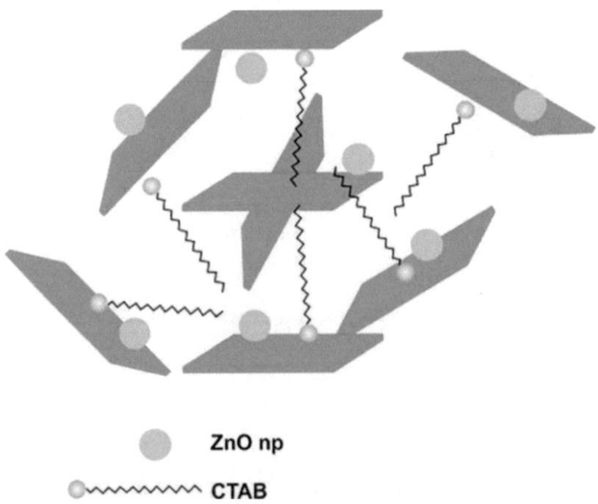

Figure 4.
Graphical illustration of MMT-CTAB/ZnO nanocomposite.

MMT was modified with CTAB to increase its interlayer distance. Subsequently, ZnO nanoparticles are incorporated onto the surface of MMT/CTAB. Analysis of XRD, TEM, SEM, and Brunauer Emmett Teller (BET) results revealed an exfoliated structure and an increase in surface area, offering more adsorption sites. The as-prepared material exhibited a notable degradation activity, resulting in a 62% yield in degradation of methylene blue [18]. In **Figure 4**, a graphical illustration of MMT-CTAB/ZnO nanocomposite.

12. MMT/TiO_2-ZnO nanocomposite for a sophisticated detection system for nanomolar level of the food preservative nitrite in sugar byproducts

An additional important application for MMT intercalated catalysis lies in the determination and surveillance of toxic elements. In this study, TiO_2 is employed as a catalysis for NiO_2^- detection, but it can be susceptible to the recombination of electron and hole, thereby affecting the electrochemical and conductivity properties of TiO_2. To address this issue, the addition of ZnO nps is necessary to improve charge transfer and catalytic efficiency. Furthermore, to limit nps agglomeration, which would decrease surface area, the addition of MMT was crucial. MMT also provides an additional adsorption site. This modification serves to immobilize ZnO nps and prevent their agglomeration. **Figure 5** displays a graphical representation of MMT-TiO_2/ZnO nanocomposite. The obtained sensor reveals a clear limit of detection at 0.12 nM for NO_2^- and it exhibited a sensitivity of 0.78 μM^{-1}. The sensor reveals satisfactory precision and strong selectivity in the detection of NO_2^- in commercial sugar byproduct samples without interference from the product's constituents [19]. **Table 5** presents previous studies that have been utilized for the detection of NO_2^- exhibiting a low limit of detection (LOD) when compared to MMT-TiO_2/ZnO.

13. Conclusions and outlook

To wrap up, MMT is an important clay mineral that has received a lot of interest for its utilization, especially in environmental fields. This chapter draws attention to the properties of MMT, its compatibility with organic compounds, such as octadecylammoinum cation and porous starch biopolymer, and their application in the adsorption of methyl orange. It is mentioned here that the structure of MMT is based on an aluminum octahedral situated between two silica tetrahedral layers. Then, it was pointed out that there are many types of MMT that are differentiated in terms of their chemical composition, geological origin, purity, and impurities. Subsequently, this chapter highlighted the interesting properties of MMT such as high isomorphic substitution and cationic exchange capacity, which are measured in the range of 80–150 meq/100 g of MMT. After that, it was pointed out that the hydratation and dehydratation of MMT varied according to the cation exchange in the interlayer space. It was found that the number of water layers estimated more than three layers for the MMT-Na. Another important property of MMT is the high surface area of MMT for about 800 m^2/g, which is initiated through three current methods: methylene blue adsorption, ethylene glycol adsorption, and N_2 adsorption/desorption (BET). All these properties ensure that this clay mineral is highly adaptable for adsorption applications.

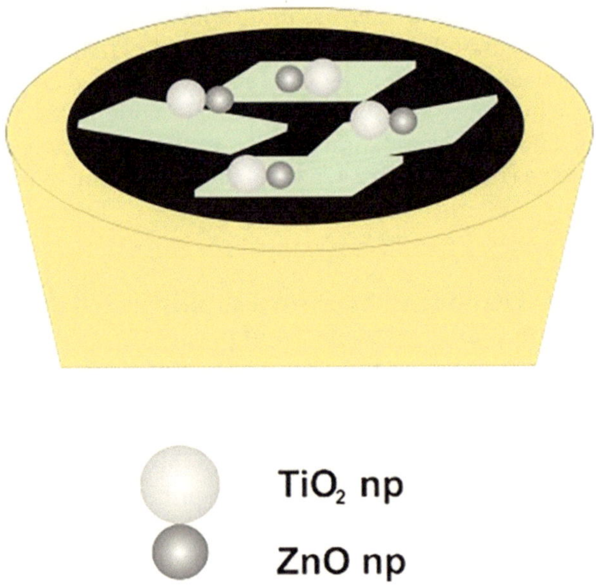

Figure 5.
Graphical representation of MMT-TiO_2/ZnO nanocomposite.

Sensor	LOD	References
Chitosan (CS)/carbon paste electrode (CPE)	0.187 µg/ml	[59]
AgPs-(ionic liquid)IL/CPE	3 mM	[60]
Ag/Cu/MWNTs	0.2 mM	[61]
(Graphene oxide)GO/SDBS/GCPE	1.89 mM	[62]

Table 5.
Determination of NiO_2^- with various sensors illustrated in the literature.

Thus, we mentioned the application of the adsorption of the nanocomposite MMT modified with octadecylammonium/porous starch (PSA/OMMT). This work merged the high properties of MMT and the porous starch biopolymer, which decreased MMT's hydrophilicity and enhanced its adsorption capacity. Indeed, it was found significant adsorption of methyl orange ability as high as 344.7 mg/g, which made that PSA/OMMT, which synthesis essentially by MMT clay mineral, is suitable as an economical and effective adsorbent for treating dyeing wastewater. Furthermore, we have studied its effectiveness in removing the ibuprofen drug by modifying it with CTAB/ZnO np. Then, we explored its potential application in the electrochemical detection of NiO_2^- by incorporating TiO_2 and ZnO nps onto MMT. Thus, we find an impressive limit of detection equal to 0.12 nM.

It is interesting to mention that MMT and some MMT-based nanocomposites can be influenced by various environmental conditions such as temperature, humidity, and acidity. Therefore, at high temperatures, they may cease to effectively adsorb or photodegrade contaminants in water. On the other hand, they can degrade over time, reducing their durability and potentially releasing contaminants into the environment upon degradation. To overcome these challenges, it is necessary to optimize their performance. MMT can be modified with non-toxic compounds, such as biopolymers, to enhance its thermal, mechanical, and durability properties, thereby prolonging the duration of adsorption and photodegradation of contaminants and ensuring biodegradation without contaminating the soil over time. Additionally, research into efficient and low-cost materials for industrial-scale applications is essential.

Acknowledgements

I want to thank Mr. Abdesslem Ben Haj Amara and Ms. Hafsia Ben Rhaiem for their invaluable assistance in providing necessary documents and correcting the manuscript and their continuous support. Thanks to Dr. Mohamed Amine Djebbi for his constant support and guidance. I am also grateful to Ramzi Chalghaf for his help. We appreciate all those who contributed, whether directly or indirectly, to the development of this work.

Conflict of interest

The authors declare no conflict of interest.

Author details

Rihem Jemai*, Ramzi Chalghaf, Saber Boubakri, Mohamed Amine Djebbi, Sonia Naamen, Hafsia Ben Rhaiem and Abdesslem Ben Haj Amara
Faculty of Sciences of Bizerte, Laboratory of Ressources, Materials and Ecosystem (RME), University of Carthage, Zarzouna, Tunisia

*Address all correspondence to: rihemdjemi@hotmail.fr

References

[1] Park J-H, Shin H-J, Kim MH, Kim J-S, Kang N, Lee J-Y, et al. Application of montmorillonite in bentonite as a pharmaceutical excipient in drug delivery systems. Journal of Pharmaceutical Investigation. 2016;**46**:363-375. DOI: 10.1007/s40005-016-0258-8

[2] Biglari Quchan Atigh Z, Sardari P, Moghiseh E, Asgari Lajayer B, Hursthouse AS. Purified montmorillonite as a nano-adsorbent of potentially toxic elements from environment: An overview. Nanotechnology for Environmental Engineering. 2021;**6**:12

[3] De Pablo L, Chávez ML, Abatal M. Adsorption of heavy metals in acid to alkaline environments by montmorillonite and Ca-montmorillonite. Chemical Engineering Journal. 2011;**171**:1276-1286. DOI: 10.1016/j.cej.2011.05.055

[4] Hang PT. Methylene blue absorption by clay minerals. Determination of surface areas and cation exchange capacities (clay-organic studies XVIII). Clays and Clay Minerals. 1970;**18**:203-212. DOI: 10.1346/CCMN.1970.0180404

[5] Kaur M, Datta M. Diclofenac sodium adsorption onto montmorillonite: Adsorption equilibrium studies and drug release kinetics. Adsorption Science & Technology. 2014;**32**:365-387. DOI: 10.1260/0263-6174.32.5.365

[6] Abdalqadir M, Rezaei Gomari S, Pak T, Hughes D, Shwan D. A comparative study of acid-activated non-expandable kaolinite and expandable montmorillonite for their CO_2 sequestration capacity. Reaction Kinetics, Mechanisms, and Catalysis. 2023;**137**:375-398. DOI: 10.1007/s11144-023-02521-w

[7] Ravichandran J. Properties and catalytic activity of acid-modified montmorillonite and vermiculite. Clays and Clay Minerals. 1997;**45**:854-858. DOI: 10.1346/CCMN.1997.0450609

[8] Bhattacharyya KG, Gupta SS. Adsorptive accumulation of Cd(II), Co(II), Cu(II), Pb(II), and Ni(II) from water on montmorillonite: Influence of acid activation. Journal of Colloid and Interface Science. 2007;**310**:411-424. DOI: 10.1016/j.jcis.2007.01.080

[9] Fatimah I, Citradewi PW, Fadillah G, Sahroni I, Purwiandono G, Dong R. Enhanced performance of magnetic montmorillonite nanocomposite as adsorbent for Cu(II) by hydrothermal synthesis. Journal of Environmental Chemical Engineering. 2021;**9**:104968. DOI: 10.1016/j.jece.2020.104968

[10] Varadwaj GBB, Parida KM. Montmorillonite supported metal nanoparticles: An update on syntheses and applications. RSC Advances. 2013;**3**:13583. DOI: 10.1039/c3ra40520f

[11] Kumar JP, Ramacharyulu PVRK, Prasad GK, Singh B. Montmorillonites supported with metal oxide nanoparticles for decontamination of sulfur mustard. Applied Clay Science. 2015;**116-117**:263-272. DOI: 10.1016/j.clay.2015.04.007

[12] Cui R, Narayanan Nair AK, Yang Y, Sun S. Molecular simulation study of montmorillonite in contact with ethanol. Industrial and Engineering Chemistry Research. 2023;**62**:2978-2988. DOI: org/10.1021/acs.iecr.2c03903

[13] Gopakumar TG, Lee JA, Kontopoulou M, Parent JS. Influence of clay exfoliation on the physical properties

of montmorillonite/polyethylene composites. Polymer. 2002;**43**:5483-5491. DOI: 10.1016/S0032-3861(02)00403-2

[14] Moet AS, Akelah A. Polymer-clay nanocomposites: Polystyrene grafted onto montmorillonite interlayers. Materials Letters. 1993;**18**:97-102. DOI: 10.1016/0167-577X(93)90064-5

[15] An J-H, Dultz S. Adsorption of tannic acid on chitosan-montmorillonite as a function of pH and surface charge properties. Applied Clay Science. 2007;**36**:256-264. DOI: 10.1016/j.clay.2006.11.001

[16] Qiu J, Cui K, Wu P, Chen G, Wang Y, Liu D, et al. The adsorption characteristics and mechanism of montmorillonite with different layer charge density for alkyl ammonium with different carbon chain length. New Journal of Chemistry. 2021;**45**:10331-10339. DOI: 10.1039/D1NJ01683K

[17] Jemai R, Djebbi MA, Boubakri S, Ben Rhaiem H, Amara BH, A. Effective removal of methyl orange dyes using an adsorbent prepared from porous starch aerogel and organoclay. Colorants. 2023;**2**:209-229. DOI: 10.3390/colorants2020014

[18] Akkari M, Aranda P, Ben Haj Amara A, Ruiz-Hitzky E. Organoclay hybrid materials as precursors of porous ZnO/silica-clay heterostructures for photocatalytic applications. Beilstein Journal of Nanotechnology. 2016;**7**:1971-1982. DOI: 10.3762/bjnano.7.188

[19] Elfiky M, Salahuddin N. Advanced sensing platform for nanomolar detection of food preservative nitrite in sugar byproducts based on 3D mesoporous nanorods of montmorillonite/TiO_2–ZnO hybrids. Microchemical Journal. 2021;**170**:106582. DOI: 10.1016/j.microc.2021.106582

[20] Grim RE. Clay Mineralogy. 2nd ed. New York: McGraw-Hill; 1968

[21] Jmaïel Ben B. Contribution A L'EtudeDes Systemes Eau-Argile ParDiffusion Des Rayons X – Structure Des Couches Inserees Et Mode D'Empilement Des Feuillets Dans Les Hydrates Homogenes A Une Et Deux Couches D'eaux De La Beidellite-Na. Faculty of sciences of Tunisia, Manar university; 1985

[22] Veiskarami M, Sarvi MN, Mokhtari AR. Influence of the purity of montmorillonite on its surface modification with an alkyl-ammonium salt. Applied Clay Science. 2016;**120**:111-120. DOI: 10.1016/j.clay.2015.10.026

[23] Ben Rhaien H, Tessier D, Pons CH. Comportement hydrique et évolution structurale et texturale des montmorillonites Au tours d'un cycle de dessiccation-humectation: Partie I. Cas Des Montmorillonites Calciques. Clay Minerals. 1986;**21**:9-29. DOI: 10.1180/claymin.1986.021.1.02

[24] Chalghaf R. Détermination des propriétés structurales d'un phyllosilicate 2:1 Tunisien de la région de Gafsa. Tunisia: Faculty of Sciences of Bizerte/University of Carthage; 2009

[25] Bergaya F, Theng BKG, Lagaly G, editors. Handbook of Clay Science. Developments in Clay Science. Orléans, France: CRMD, CNRS-Université d'Orléans; 2006

[26] Uddin F. Montmorillonite: An introduction to properties and utilization. In: Zoveidavianpoor M, editor. Current Topics in the Utilization of Clay in Industrial and Medical Applications. London, UK: InTech; 2018. DOI: 10.5772/intechopen.77987

[27] He H, Guo J, Xie X, Lin H, Li L. A microstructural study of acid-activated

montmorillonite from Choushan, China. Clay Minerals. 2002;**37**:337-344. DOI: 10.1180/0009855023720037

[28] Hoffman V, Klemen R. Loss on heating of the ability of lithium ions to exchange in bentonite. Zeitschrift für Anorganische und Allgemeine Chemie. 1950;**262**:95-99

[29] Jr BAJ, Broad WC, Flaschka H. The ethylenediaminetetraacetic acid (EDTA) titration: Nature and methods of end-point detection (1). Chemist-Analyst. 1956;**45**(86-93):111-112

[30] Jr BAJ, Broad WC, Flaschka H. The ethylenediaminetetraacetic acid (EDTA) titration: Nature and methods of end-point detection (II). Chemist-Analyst. 1957;**46**:18-28

[31] Karakassides MA, Petridis D, Gournis D. Infrared reflectance study of thermally treated Li- and Cs-montmorillonites. Clays and Clay Minerals. 1997;**5**:649-658. DOI: 10.1346/CCMN.1997.0450504

[32] Dejou J. La surface spécifique des argiles, sa mesure, relation avec la CEC et son importance agronomique. La capacité d'échange cationique et la fertilisation des sols. 1987:72-83

[33] Chalghaf R. Etude des propriétés d'hydratation d'une smectite de référence soumise à l'effet de contrainte de température, de pH et d'humidité relative par simulation des raies de diffraction: Application à la sélectivité. Tunisia: Faculty of Sciences of Bizerte/University of Carthage; 2014

[34] Ben Rhaiem H. Etude du comportement hydrique des montmorillonites calciques et sodiques par analyse de la Diffusion Des Rayons X Aux Petits Angles: Mise En Evidence De la Transition Solide Hydraté. Tunisia: Faculty of Sciences of Bizerte/ university of Carthage

[35] Shibata T, Fukuzumi Y, Kobayashi W, Moritomo Y. Fast discharge process of layered cobalt oxides due to high Na^+ diffusion. Scientific Reports. 2015;**5**:9006. DOI: 10.1038/srep09006

[36] Chalghaf R, Oueslati W, Ammar M, Ben Rhaiem H, Abdesslem BHA. Effect of temperature and pH value on cation exchange performance of a natural clay for selective (Cu^{2+}, Co^{2+}) removal: Equilibrium, sorption and kinetics. Progress in Natural Science: Materials International. 2013;**23**:23-35. DOI: 10.1016/j.pnsc.2013.01.004

[37] Ammar M, Oueslati W, Chorfi N, Ben Rhaiem H. Interlamellar space configuration under variable environmental conditions in the case of Ni-exchanged montmorillonite: Quantitative XRD analysis. Journal of Nanomaterials. 2014;**2014**:1-13. DOI: 10.1155/2014/284612

[38] Ben Rhaiem H. Analyse multiéchelle de phyllosilicates tunisiens par diffraction et par diffusion aux petits angles des R. X. et par MET. Relation entre structure, microteture et propriétés macroscopiques d'hydratation au cours d'un cycle de dessication-humectation. 1999

[39] Greene-Kelly R. Dehydration of the montmorillonite minerals. Mineralogical Magazine and Journal of the Mineralogical Society. 1955;**30**:604-615. DOI: 10.1180/minmag.1955.030.228.06

[40] Lint-Nerized RJP, Starches. Gel filtration and enzymatic studies of insoluble residues from prolonged acid treatment of potato starch. Cereal Chemistry. 1974;**51**:389-406

[41] Hu Y, Song L, Xu J, Yang L, Chen Z, Fan W. Synthesis of polyurethane/clay

intercalated nanocomposites. Colloid and Polymer Science. 2001;**279**:819-822

[42] Zilg C, Thomann R, Mülhaupt R, Finter J. Polyurethane nanocomposites containing laminated anisotropic nanoparticles derived from organophilic layered silicates. Advanced Materials. 1999;**11**:49-52

[43] Umpuch C, Sakaew S. Removal of methyl orange from aqueous solutions by adsorption using chitosan intercalated montmorillonite. Songklanakarin Journal of Science & Technology. 2013;**4**:35

[44] Mahdavinia GR, Aghaie H, Sheykhloie H, Vardini MT, Etemadi H. Synthesis of CarAlg/MMt nanocomposite hydrogels and adsorption of cationic crystal violet. Carbohydrate Polymers. 2013;**98**:358-365. DOI: 10.1016/j.carbpol.2013.05.096

[45] Garcia-Padilla A, Moreno-Sader KA, Realpe A, Acevedo-Morantes M, Soares JB. Evaluation of adsorption capacities of nanocomposites prepared from bean starch and montmorillonite. Sustainable Chemistry and Pharmacy. 2020;**17**:100292

[46] Dao TBT, Ha TTL, Nguyen TD, Le HN, Ha-Thuc CN, Nguyen TML, et al. Effectiveness of photocatalysis of MMT-supported TiO_2 and TiO_2 nanotubes for rhodamine B degradation. Chemosphere. 2021;**280**:130802

[47] Bouwe RGB, Tonle IK, Letaief S, Ngameni E, Detellier C. Structural characterisation of 1,10-phenanthroline–montmorillonite intercalation compounds and their application as low-cost electrochemical sensors for Pb(II) detection at the sub-nanomolar level. Applied Clay Science. 2011;**52**:258-265. DOI: 10.1016/j.clay.2011.02.028

[48] Elfiky M, Salahuddin N, Matsuda A. Green fabrication of 3D hierarchical blossom-like hybrid of peeled montmorillonite-ZnO for in-vitro electrochemical sensing of diltiazem hydrochloride drug. Materials Science and Engineering: C. 2020;**111**:110773. DOI: 10.1016/j.msec.2020.110773

[49] Johnson CE Jr. Methylene blue adsorption and surface area measurements. In: 131st National Meeting of the American Chemical Society. Washington, DC: American Chemical Society; 1957. pp. 7-12

[50] Dogan AU, Dogan M, Onal M, Sarikaya Y, Aburub A, Wurster DE. Baseline studies of the clay minerals society source clays: Specific surface area by the Brunauer Emmett Teller (BET) method. Clays and Clay Minerals. 2006;**54**:62-66. DOI: 10.1346/CCMN.2006.0540108

[51] Dyal RS, Hendricks SB. Total surface of clays in polar liquids As A characteristic index. Soil Science. 1950;**69**:503-509

[52] Oleyaei SA, Almasi H, Ghanbarzadeh B, Moayedi AA. Synergistic reinforcing effect of TiO_2 and montmorillonite on potato starch nanocomposite films: Thermal, mechanical and barrier properties. Carbohydrate Polymers. 2016;**152**:253-262. DOI: 10.1016/j.carbpol.2016.07.040

[53] Zhu J, Zhang S, Pu H, Chen X, Zou S, Li L, et al. Structural properties of propionylated starch-based nanocomposites containing different amylose contents. International Journal of Biological Macromolecules. 2020;**149**:532-540. DOI: 10.1016/j.ijbiomac.2020.01.274

[54] Atta AM, Al-Lohedan HA, Alothman ZA, Abdel-Khalek AA, Tawfeek AM. Characterization of reactive amphiphilic montmorillonite

nanogels and its application for removal of toxic cationic dye and heavy metals water pollutants. Journal of Industrial and Engineering Chemistry. 2015;**31**:374-384. DOI: 10.1016/j.jiec.2015.07.012

[55] Wilpiszewska K, Antosik AK, Spychaj T. Novel hydrophilic carboxymethyl starch/montmorillonite nanocomposite films. Carbohydrate Polymers. 2015;**128**:82-89. DOI: 10.1016/j.carbpol.2015.04.023

[56] Ai Z, Liu C, Zhang Q, Qu J, Li Z, He X. Adding ZnO and SiO_2 to scatter the agglomeration of mechanochemically prepared Zn-Al LDH precursor and promote its adsorption toward methyl orange. Journal of Alloys and Compounds. 2018;**763**:342-348

[57] Ke P, Zeng D, Xu K, Cui J, Li X, Wang G. Preparation of quaternary ammonium salt-modified chitosan microspheres and their application in dyeing wastewater treatment. ACS Omega. 2020;**5**:24700-24707

[58] Mohammadi N, Khani H, Gupta VK, Amereh E, Agarwal S. Adsorption process of methyl orange dye onto mesoporous carbon material–kinetic and thermodynamic studies. Journal of Colloid and Interface Science. 2011;**362**:457-462

[59] Habib IH. Anodic stripping voltammetric determination of nitrite using carbon paste electrode modified with chitosan. American Journal of Analytical Chemistry. 2011;**2**:284

[60] Menart E, Jovanovski V, Hoˇcevar SB. Silver particle-decorated carbon paste electrode based on ionic liquid for improved determination of nitrite. Electrochemistry Communications. 2015;**52**:45-48

[61] Zhang Y, Nie J, Wei H, Xu H, Wang Q, Cong Y, et al. Electrochemical detection of nitrite ions using Ag/Cu/MWNT nanoclusters electrodeposited on a glassy carbon electrode. Sensors & Actuators. 2018;**258**:1107-1116

[62] Berisha L, Maloku A, Haliti M, Jashari G, Ukmata A, Sýs M. Voltammetric determination of nitrites in meat products after reaction with ranitidine producing 2-methylfuran cation. Microchemical Journal. 2020;**159**:105403. DOI: 10.1016/j.microc.2020.105403

Chapter 3

Effects of Mixing Sequence on the Morphological and Rheological Properties of Polymer-Clay Nanocomposites of Poly(Butylene Adipate-co-Terephthalate) and Poly(Lactic Acid) Blend

Mahin Shahlari, Bahareh Baheri and Sunggyu Lee

Abstract

The effect of sequential mixing of poly(butylene adipate-co-terephthalate) (PBAT) and poly(lactic acid) (PLA) with organically modified silicate layers on the dispersion of the organoclay particles and its effect on the morphology of the blended polymer clay nanocomposite (BPCN) was examined. The dispersion of the organoclay platelets, morphology of the blend composite, and rheological and thermomechanical properties of these composites with 50/50 ratio of PLA to PBAT were investigated using X-Ray diffraction analysis, scanning and transmission electron microscopy, dynamic mechanical shear test, and thermogravimetric analysis. The morphology of blends with 70/30 ratio of PLA to PBAT was also examined and compared. The sequential mixing of the organoclay with this BPCN enhanced the level of clay dispersion when compared to the simultaneously blended nanocomposites. Larger PLA domains and better clay dispersion in PLA phase were observed when clay was first mixed with PLA and then subsequently mixed with PBAT.

Keywords: nanocomposite, compatibilized blend, PBAT, PLA, BPCN, property characterization

1. Introduction

In response to the escalating global demand for sustainable alternatives and the imperative to reduce the environmental impact of traditional commodity plastics, there has been a pronounced surge in interest in biodegradable polymers. These polymers, designed to naturally break down over time, offer a promising solution to mitigate the persistent environmental issues associated with nondegradable plastics. As industries and consumers increasingly recognize the importance of eco-friendly materials, there is a compelling need and motivation to further develop and refine the properties of existing biodegradable polymers. Traditional plastics, notorious for their prolonged persistence in the environment, have triggered a paradigm shift in consumer preferences and industry practices. Biodegradable polymers present a sustainable pathway forward, but for their widespread adoption, they must surpass the performance benchmarks set by conventional plastics [1, 2].

Among the extensively studied biodegradable polymers, poly(lactic acid) (PLA) has gained significant attention due to its production from renewable resources such as corn-starch, sugar, and wheat, along with its relatively high modulus [3–8]. However, PLA is hindered by brittleness and a lack of flexibility. To address these shortcomings, blending PLA with other biodegradable polymers has been explored as a viable solution [9–12].

Poly(butylene adipate-co-terephthalate) (PBAT), a biodegradable polymer with high elasticity, is one of the polymers whose blend with PLA has been studied by several researchers [13–16]. The resulting blends of PLA and PBAT exhibited an increased elongation at break compared to PLA alone. Nevertheless, the blends demonstrated lower values for both modulus and strength in comparison to pure PLA [14]. Blends of unmodified PLA and PBAT are poorly miscible without an aid of a compatibilizing effect, thus exhibiting a tendency of macro-separation. Properly functionalized nanoparticles such as organically modified nanoclay could provide a needed compatibilizing effect for blends of PLA and PBAT while rendering reinforcements of polymeric matrix.

Incorporating functional nanoparticles, a well-documented method for fortifying the modulus and strength of polymers [3, 17–19], including PLA [20], emerges as a strategic intervention. Therefore, to regain a high modulus for the PLA composite despite the softening effect of PBAT addition, nanoparticles have been compounded with PLA and PBAT blends [21–24].

Numerous potential applications for PBAT could find advantages in utilizing PBAT and PLA blend, resulting in an enhanced modulus and renewable content. Introducing nanoparticles not only serves to increase the blend's modulus but also has the potential to increase the heat deflection temperature, thermomechanical properties, and barrier properties of the blend. Organoclay has been found to be effective to improve these properties in PLA composites [25–29].

In composites of polymer blends, the location of the nanoparticles can impart a significant effect on the final properties, therefore controlling the location of these particles in a blend can be a great tool in tailoring polymers with desired properties.

Principally, the dispersed location of organoclay depends on the interaction between the two polymers as well as the interaction of each polymer with the organoclay particles. However, the high viscosity of the polymers does not allow the systems to always reach a thermodynamic equilibrium state, and the kinetic factors involved in the mixing process could influence and determine the location of the particles in the polymer. When the polymers and the particles are added simultaneously, regardless of the real affinity between the polymer and the organoclay, the polymer with a lower melting temperature typically exhibits better interaction with particles [30]. Therefore, the mixing sequence of the components could have a significant effect on how organoclay platelets are dispersed in the blend and ultimately affect the characteristics of the blend [31].

The effects of the mixing sequence on the properties of blends have been different from case to case. Kim et al. showed an increased degree of exfoliation of the clay when poly(ε-caprolactone) was initially mixed with clay and then blended with poly(styrene-co-acrylonitrile) [32]. Increased heat deflection temperature and flexural modulus were reported when ethylene vinyl acetate (EVA) was mixed with organoclay and Cloisite 20A and then mixed with propylene-(ethylene-propylene) copolymer (PP–EP) [33]. Dasari et al. observed that for a blend of Nylon66 and maleated styrene-ethylene/butylene-styrene block copolymer (SEBS-g-MA), the best impact strength was achieved when Cloisite 30B was mixed with Nylon66 first [34]. Tao et al. studied the migration behavior of carbon nanotubes (CNTs) within a co-continuous poly(lactic acid)/poly(ε-caprolactone) (PLA/PCL) blend when pre-compounding CNTs in one phase and controlling the mixing time. Enhanced viscosity compatibility between the two polymer phases led to a reduction in the size of the respective phases. Specifically, composites subjected to premixing of PCL/CNT followed by the introduction of PLA under low shear stress conditions demonstrated a reduction in the size of the individual phases [35].

On the other hand, some studies showed better properties were achieved by the simultaneously blended composites [36, 37]. Vo et al. reported the highest mechanical properties for Cloisite 30B simultaneously compounded with polyvinylidenefluoride (PVDF) and nylon 6 with Cloisite 30B mostly dispersed in Nylon 6 and some located at interface. The location of the particles was claimed to be responsible for inhibited coalescence of the dispersed phase [36]. In some studies, the mixing order did not cause a significant effect [38–40].

To investigate the interaction of the organically modified clays with the PBAT and PLA blends and to study the effect of the mixing sequence on the dispersed location of the nanoclay particles, a series of experiments were designed that considered the mixing sequence to be a main factor. Clay was first mixed with either PBAT or PLA; then, the second polymer was blended with the compound. In one set of samples, clay was dry-mixed with the two polymers prior to melt-mixing. The morphology and rheological properties of the blends were analyzed to determine any significant changes that may occur because of varying the mixing order. The transition of the particles from one polymer to the other was studied as well.

2. Experimental

2.1 Materials

The PLA used in this study was a product of NatureWorks LLC (United States), PLA 4042, with a density of 1.24 g/cm^3, glass transition temperature of 52–58°C, and melting temperature of 150°C. The structure of this polymer is shown in **Figure 1**.

The PBAT was a BASF product marketed as Ecoflex® FBX 7011, with a mass density of 1.25–1.27 g/cm^3, glass transition temperature of about −29°C, melting temperature of 115°C, and melt flow rate of 3.3–6.6 g/10 min at 190°C and 2.16 Kg [41]. The structure of this polymer is shown in **Figure 2**.

The organically modified nanoclay, Cloisite 30B, was supplied by Southern Clay Products Corporation. Cloisite 30B is montmorillonite (MMT)-type clay. Based on the technical data sheets of Southern Clay Products, Inc., Cloisite 30B was modified by methyl, tallow, bis-2-hydroxyethyl, and quaternary ammoniums. The amount of surfactant for Cloisite 30B is 90 meq/100 g. Tallow is a mixture of octadecyl, hexadecyl, and tetradecyl with octadecyl as the major component [42].

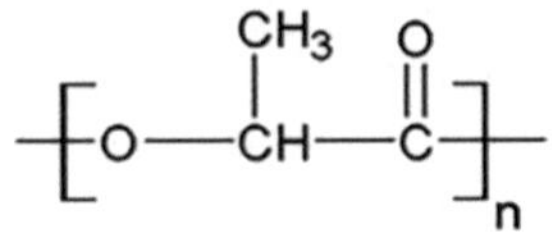

Figure 1.
PLA structure.

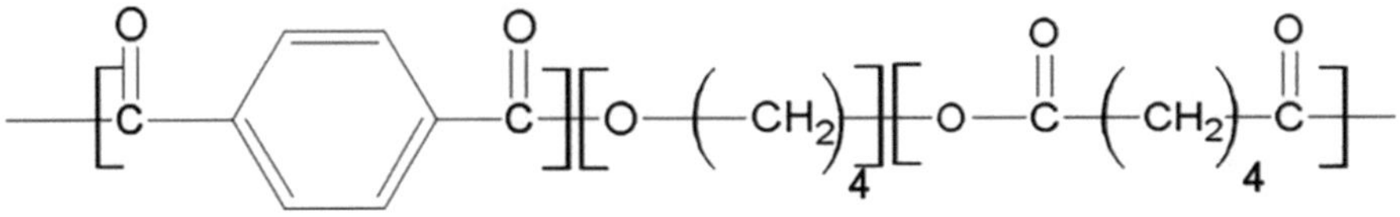

Figure 2.
PBAT structure.

2.2 Sample preparations

Both the polymer pellets and the organoclay particles were dried in a vacuum oven for 12 hours at 65°C to reduce the chance of hydrolytic degradation during processing. The compounds were prepared in a Rheomix batch mixer (Haake Inc.) at a constant speed of 30 rpm at 190°C.

For one set of samples, clay was mixed with PBAT or PLA for 5 minutes before the second polymer was added and melt-blended for an additional 5 minutes. Another set was prepared by mixing clay for 10 minutes with the melt of the two polymers, which had been dry-mixed prior to melt mixing. The PBAT/PLA ratios considered here were 50/50 and 70/30. This work carried out the sequential mixing in a batch system; however, the process is transferable to an extrusion system by first compounding each polymer separately with organoclay and then blending the pelletized composite with the neat blend in the extruder.

The samples are named by the ratio of the first polymer mixed with organoclay followed by the letter B representing Cloisite 30B and its weight percentage in the blend. Therefore, 50PLA/B3 would be a blend in which the ratio of PLA to PBAT is 50/50 and PLA is first mixed with 3 wt.% Cloisite 30B and 70PLA/B3 would be a blend with the ratio of PLA to PBAT of 70/30 with similar clay content and mixing procedure.

2.3 Characterization

The X-ray Diffraction (XRD) analyses were carried out using a Philips X-Pert diffractometer at ambient temperature with Cu Kα radiation with wavelength of 0.154 nm, voltage of 45 kV, and current of 40 mA. The scanning step was 0.03 degrees, and the time per step was 10 seconds. The samples used for XRD were molded using a hydraulic hot-press. The interlayer spacing (d-spacing) of each sample was determined from the angular location of the peaks caused by the basal reflection of the repetitive 22 structure of the silicate layers using the Bragg's relationship [43]. In case of complete exfoliation of the silicate layers, no peak should be observed. However, the lack of a peak is not a certain indication of exfoliation. The shift of the characteristic peak to lower angles results in higher calculated d-spacing for the clay and is attributable to the intercalation of the silicate layers [43].

The thermomechanical properties of the molded compounds were tested using a Dynamic Mechanical Thermal Analyzer (DMTA) IV from Rheometric Scientific

Inc., and Rheometric Orchestrator software was used to operate the instrument and analyze the data. A dynamic strain sweep test was conducted prior to these tests to ensure that the applied strain was within the linear viscoelastic region for the blends. Dynamic temperature ramp tests were then conducted using a three-point bending fixture at a frequency of 1 s^{-1} and a constant strain of 0.1%. The storage and loss moduli were acquired while the specimens were heated from 30–100°C at a rate of 2°C/min. Five specimens were tested for each sample set.

Scanning Electron Microscopy (SEM) was performed using a Hitachi S4700 FESEM. The samples were immersed in liquid nitrogen and then fractured. Subsequently, they were sputter-coated with gold-palladium. An accelerating voltage of 5 kV was used to avoid charging accumulation on the sample surface, and a short working distance of 3–5 mm at ultra-high resolution mode was adopted to ensure vivid images. The lack of a selective solvent for these two polymers prevented etching the samples prior to SEM imaging [42].

Transmission Electron Microscopy (TEM) analysis was performed using a JEOL JEM-1400 with an accelerating voltage of 120KV. Specimens were cryomicrotomed to an average thickness of 70 nm. The TEM pictures were taken using a Gatan 2K×2K digital camera. The images were analyzed using Image J software. TEM images make it possible to evaluate the dispersion of silicate layers in the blend and to identify their location.

The Thermogravimetric Analysis (TGA) was performed using Netzsch STA 409C/CD with a nitrogen flow rate of 20 ml/min. The samples were heated to 500°C at a linear rate of 10°C/min. Approximately 100 mg of grinded compound was used as a sample.

3. Result and discussion

3.1 XRD analysis

An X-ray diffractometry (XRD) analysis was used to investigate the dispersion of the organoclay within the blends. The basal reflection of the repetitive structure of the silicate layers (if the latter are not completely exfoliated) creates low-angle peaks corresponding to the interlayer spacing of the silicate layers.

Figure 3 shows the XRD diffraction pattern of the sequentially mixed blends. The interlayer spacing of the samples is determined from the angular location of the peaks using the Bragg's law. The XRD peak positions and interlayer spacing of 5050premixed/B3, 50PLA/B3, and 50PBAT/B3 samples are shown in **Table 1**. If the peaks are assumed to be related to (001) plane, n = 1, the spacings for all three compounds would be lower than 1.88 nm, the original d-spacing of the Cloisite 30B. This decrease in the space between the layers, although less likely, can be attributed to the degradation of some of the methyl, tallow, bis-2-hydroxyethyl, and quaternary ammonium present as the organic modifiers in Cloisite 30B.

Sample	2θ°	Interlayer spacing(nm)
5050premixed/B3	5.271	3.35
50PLA/B3	5.167	3.42
50PBAT/B3	5.141	3.44

Table 1.
XRD peak positions and interlayer spacing of 5050premixed/B3, 50PLA/B3, and 50PBAT/B3 samples.

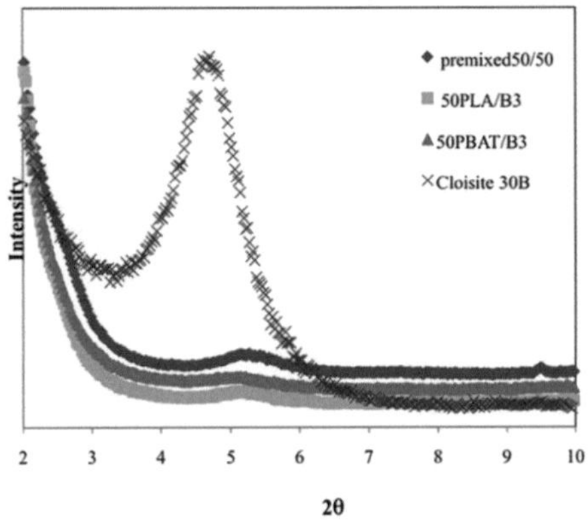

Figure 3.
XRD patterns of (a) 5050premixed/B3, (b) 50PBAT/B3, and (c) 50PLA/B3.

The interlayer spacing of the samples did not indicate a significant difference between the two sequentially mixed samples, but the premixed sample showed a smaller d-spacing compared to the other two samples. As will be explained later in the TEM analysis section, the premixed samples showed a higher content of organoclay tactoids and agglomerates. The agglomerates could locally concentrate the heat and expedite the degradation of the organic modification in the premixed blend.

3.2 Rheological analysis

The rheological behavior of a polymer system is directly affected by the microstructure and properties of the particles that compounded with the polymer. Due to this fact, it is possible to evaluate the particle's relative dispersion state in the matrix [44, 45].

As shown in **Figure 4**, at low frequencies, the shear storage modulus (G') and shear loss modulus (G") of clay-containing blends showed a large increase compared to those of the neat blend. The lowest storage and loss moduli among clay-composites were observed for the premixed samples, as the storage modulus of 50PBAT/B3 and 50PLA/B3 at lowest frequency were 205% and 123% higher than that of the premixed composite. The higher storage and loss modulus in the sequentially mixed blends could be due to the greater level of clay dispersion in this blend. The rheological properties are highly dependent on the population (i.e., number) of particles dispersed per volume of the matrix.

In general, polymers in a melt state show a storage modulus larger than their loss modulus. As shown in **Figure 4**, unlike the neat blend, the loss moduli of the clay-containing blends were lower than the storage moduli at low frequencies. In addition, at lower frequencies, the slopes of the storage moduli graphs were reduced significantly, indicating less dependence on frequency. In nanocomposites, the storage modulus higher than loss modulus, in addition to a plateau for storage modulus at low frequencies, indicates solid-like behavior [46].

The storage modulus for both sequentially mixed samples had a higher value than the loss modulus at any frequency. The 5050premixed/B3 sample showed a G'/G" crossover point indicating transition to liquid-like behavior at higher frequencies for this melt. This observation in addition to larger dependence of the storage modulus on frequency (larger slope) for the premixed blend is an indication of the lower number of particles dispersed in this blend and higher level of agglomeration as mentioned above. At high frequencies, different samples showed similar moduli due to the short response time [45].

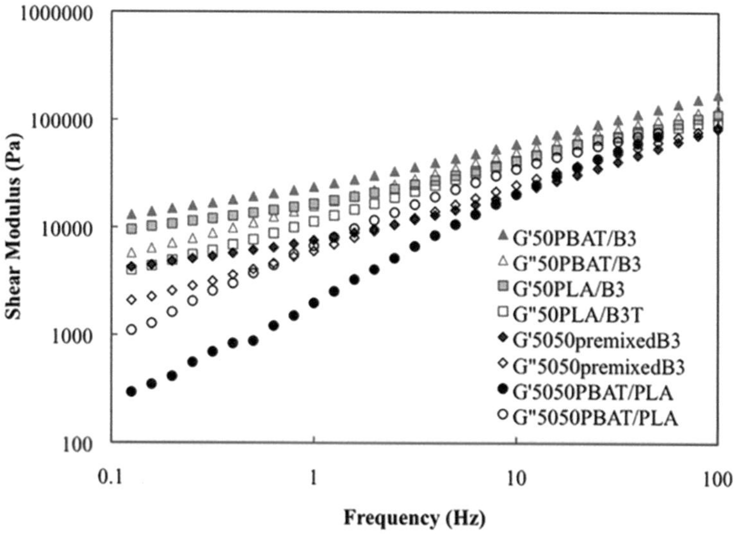

Figure 4.
Storage modulus (G') and loss modulus (G"): 5050PBAT/PLA, 5050premixed/B3, 50PBAT/B3, and 50PLA/B3 composites.

3.3 Organoclay dispersion

To determine the clay dispersion degree, and study the morphology of the blends, the TEM images of the samples were analyzed. In these images, PLA appears as the lighter phase and PBAT is the darker one (**Figure 5**). In all the blends with 50/50 ratio, exfoliated, intercalated, and agglomerated clay platelets were observed. However, in the premixed blend, large clay tactoids and agglomerates and fewer exfoliated/intercalated clay platelets were detected than the sequentially mixed samples.

The large clay tactoids observed in the premixed samples were mainly located in the PLA phase. This could be due to the low affinity of PLA to the organoclay [21]. On the other hand, since PBAT is being processed at a temperature more than 50°C above its melting temperature, it has a lower viscosity than PLA in this blend. Therefore, PBAT forms the matrix in 50/50 ratio as is generally known for polymer blends with large viscosity differences [6]. Therefore, the shear applied is significantly different for each phase. As the premixed polymers melt, some of the tactoids originally located in the PLA phase might not get exposed to a large enough shear. As PLA phase goes through successive breakups, some of the tactoids are broken and dispersed, but there is a chance that a few tactoids would be preserved in the PLA phase.

In the sequentially mixed samples, clay particles were located at the blend's interface as well as in both phases of the blend despite organoclay's initial mixing solely with one of the phases. This indicates the transition of the particles from one phase to the other.

Due to the large contrast between the two phases in the TEM images, it is difficult to detect the clay platelets in the PBAT phase, and only agglomerates are detectable in PBAT matrix, which itself is an indication of the migration of the particles form PLA phase to PBAT. Due to this limitation in addition to the key role the organoclay particles can have on the PLA phase properties, the TEM study was focused on the dispersion of the clay platelets in the PLA.

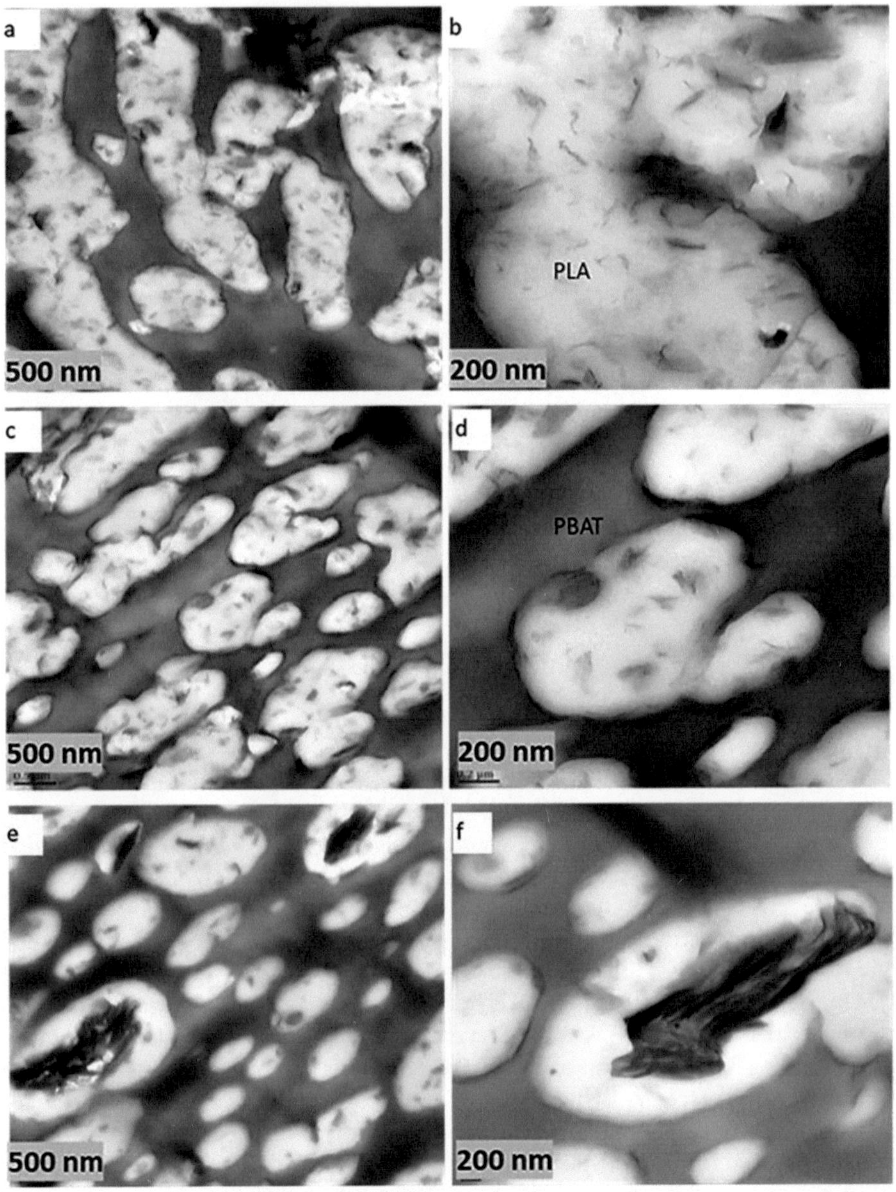

Figure 5.
TEM micrographs of the sequentially mixed blends, (a) and (b): 50PLA/B3, (c) and (d): 50PBAT/B3, and (e) and (f): 5050premixed/B3. (b), (d), and (f) images have higher magnification. The lighter phase is PLA and the darker phase is PBAT [42].

Only few groups have studied the mechanisms of particles' transition from one phase to another [30, 47]. The migration of particles from matrix to the droplets occurs through the collision of the particles with the droplets. The solidity and the small size of the particles help the drainage of the matrix polymer between the particle and the droplets. Also based on their studies, the transition of the particles from the droplets to the matrix is possible as the particles can be moved by the internal flow inside the dispersed phase to the boundary and pass through the interface if the interfacial forces are in favor for the transition. Another possibility for particles transfer is through coalescence of the dispersed phase droplets. However, this mechanism has not been studied specifically [47].

In considering the transition of the organoclay particles from PBAT to PLA droplets in the 50PBAT/B3 composites, the uneven accumulation of particles in the PLA phase could be an indication of the coalescence as the main mechanism for migration of the particles from the PBAT phase to PLA (**Figure 6**). As the PLA droplets collide, the organoclay concentrated at the interface gets trapped between the droplets and locates inside the bigger droplet that is formed (**Figure 6a**). The schematic illustration of the organoclay transfer from PBAT matrix to the PLA phase is shown in **Figure** 7.

In 50PLA/B3, since clay was first mixed with PLA, the organoclay platelets were more evenly dispersed in the PLA phase compared to the 50PBAT/B3 blends (**Figure 6(b)**). However, organoclay aggregates were observed in both phases and at the interface.

The overall better dispersion of the silicate layers in the 50PLA/B3 and 50PBAT/B3 than the premixed blend suggests that higher level of dispersion for clay can be achieved when it is mixed with a single polymer at a time than mixed with two polymers simultaneously.

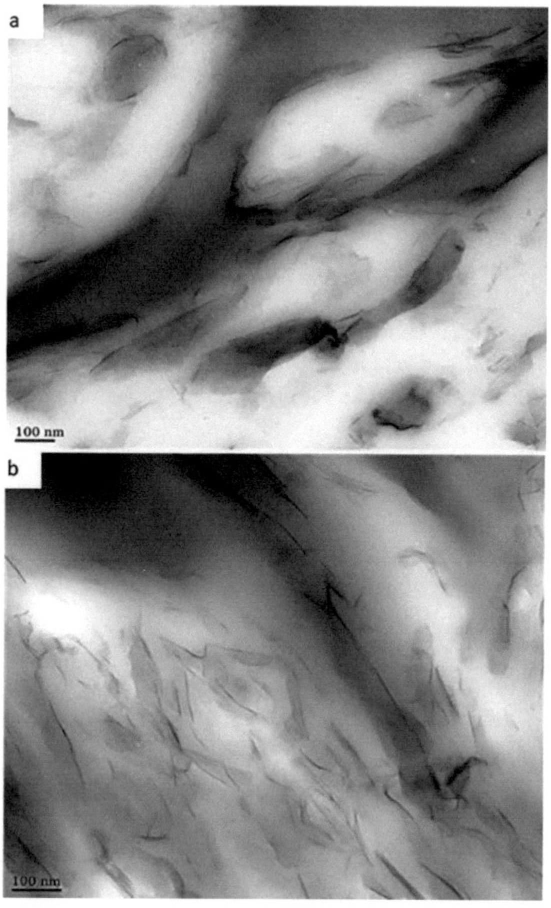

Figure 6.
Dispersion of the organoclay particles in the PLA phase. (a) 50PBAT/B3 and (b) 50PLA/B3. The lighter phase is PLA and the darker phase is PBAT [42].

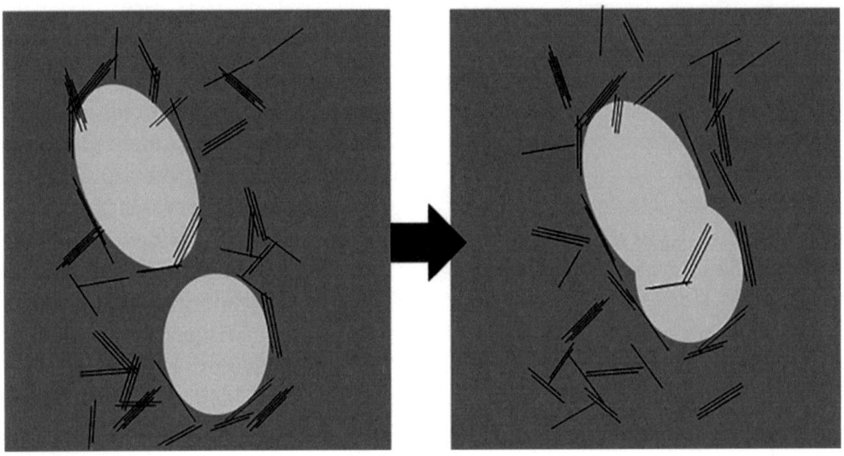

Figure 7.
Schematic illustration of organoclay transfer from the PBAT matrix to the PLA dispersed phase.

3.4 Morphology

To compare the dispersed phase sizes of the three samples, the area of the PLA domains was measured, and the measurements were divided into 4 ranges. The fraction of area in each range was calculated in order to better illustrate the size distribution of the PLA domains in each sample. These data are reported in **Table 2**, and the size distribution of the PLA phase is illustrated in **Figure 8**. Among the three sets of samples, the largest PLA domains were observed in 50PLA/B3, whereas the smallest PLA phase sizes were observed for the premixed samples. The kinetic factors involved in the 50PLA/B3 system and the effect of the organoclay particles in hindering the breakup of the PLA phase resulted in large and partially continuous PLA domains. The organoclay platelets can obstruct the breakup process by creating a network structure inside the dispersed phase, which would resist breakup of the droplet [48]. The clay platelets can also reduce the mobility of the polymer chains [49].

The SEM images of the 70PLA/B3 and 30PBAT/B3 are shown in **Figure 9**. A semi-continuous structure for PLA was observed for 70PLA/B3 blend unlike the 30PBAT/B3 one. To further detect the level of co-continuity in the 70PLA/B3 blend, TEM images of the former blend were acquired, which showed the continuity of each phase to be considerable (**Figure 10**).

Since the numerical ratio of the two polymers was kept the same in the 70PLA/B3 and 30PLA/B3 samples but the continuity was not observed in the latter samples, the ratio of the two polymers is not a significant factor in forming the semi-co-continuous morphology. The effect of organoclay's presence on the morphology of the PLA

	Maximum PLA Domain (area) (μm^2)	Overall Average (μm^2)
50PLA/B3	123.94	5.26
50PBAT/B3	13.24	1.59
5050premixed/B3	9.16	0.95

Table 2.
Effect of mixing sequence on the PLA domain size. Maximum and overall average of the PLA phase area in the 50PLA/B3, 50PBAT/B3, and 5050premixed/B3 (obtained from TEM images).

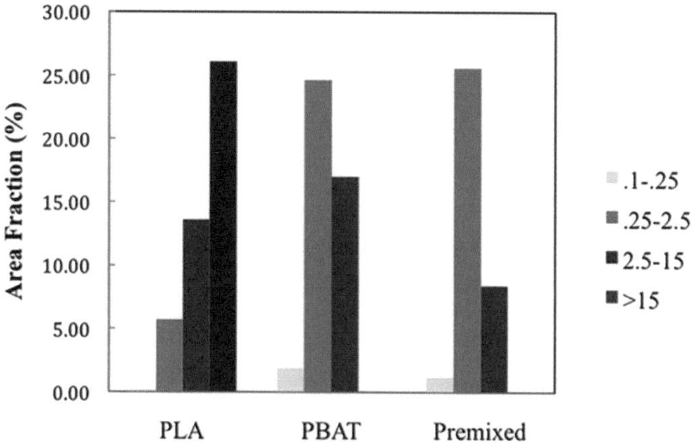

Figure 8.
Size distribution of PLA particles in the 50PLA/B3 (PLA), 50PBAT/B3 (PBAT), and 5050premixed/B3 (premixed). The units for size ranges are μm².

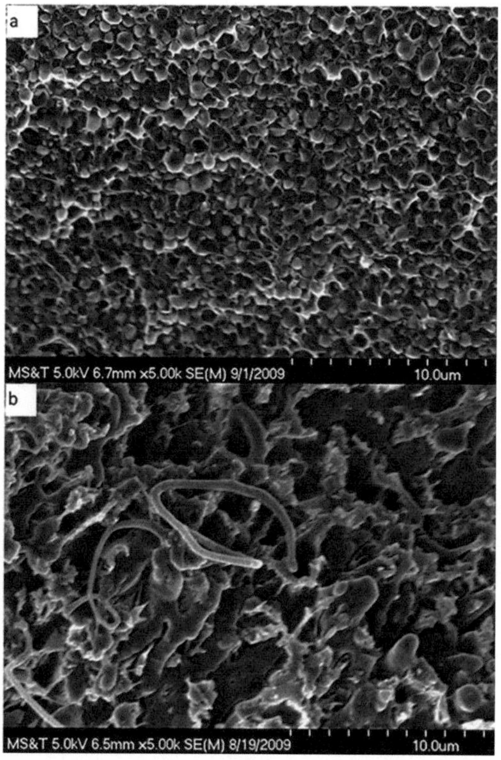

Figure 9.
SEM images of blends with PLA/PBAT ratio of 70 to 30: (a) 70PLA/B3, (b) 30PBAT/B3 [42].

phase, as was explained for the 50PLA/B3 samples, and the effect of the particles in stabilizing the kinetic factors introduced through the mixing procedure are likely the reasons for the observed continuity. **Figure 10b** shows the TEM image of 70PLA/B3 in a larger magnification, which makes it possible to observe some of the clay platelets forming a network structure in this phase. Several studies have shown formation of a

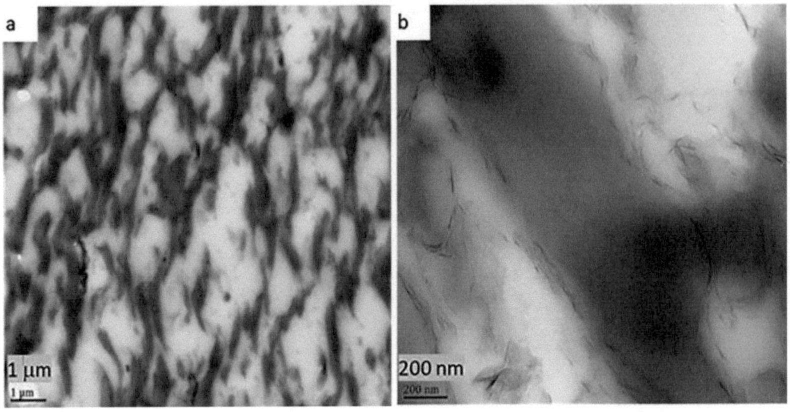

Figure 10.
TEM images of 70PLA/B3 at two magnifications. (a): 1 µm and (b) 200 nm. The lighter phase is PLA and the darker phase is PBAT [42].

co-continuous morphology with assistance of organoclay particles [48, 50]. A schematic morphology of 70PLA/B3 blend and the effect of clay network on formation of co-continuous morphology has been illustrated in **Figure 11**.

3.5 Thermal analysis

Thermal stabilities of the samples with a 50/50 ratio were analyzed using TGA instrument. **Figure 12** shows the weight loss of these samples when heated to 500°C as well as the degradation graphs of neat PLA and PBAT for comparison. As shown, the PLA degradation occurs prior to PBAT. Therefore, the temperatures reported for up to 10% weight loss in **Table 3** should be explained with regard to PLA's degradation. A 1% weight loss did not occur till the temperature reached 272.53°C for 50PLA/B3, 279.89°C for 50PBAT/B3, and 287.1°C for the 5050premixed/B3 sample.

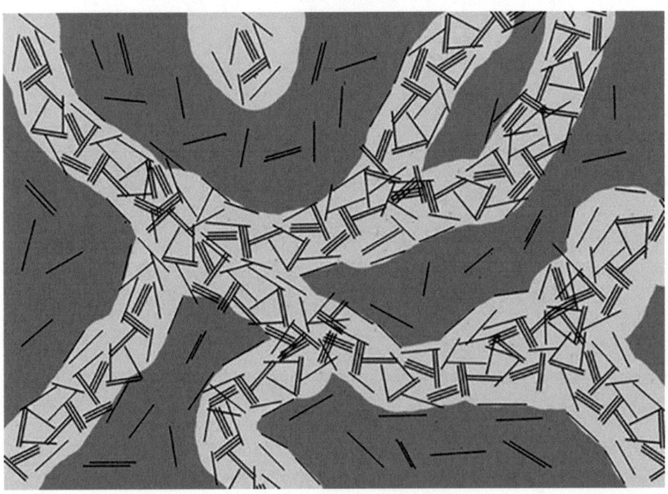

Figure 11.
Schematic illustration of the effect of clay network on the morphology of 70PLA/B3 sample. Organoclay network enhances the continuity of the phase. The lighter phase is PLA and the darker phase is PBAT.

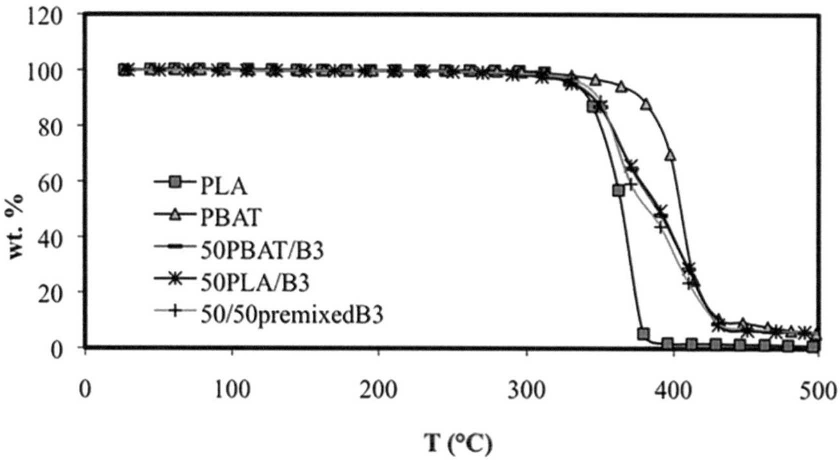

Figure 12.
Thermal degradation of PLA, PBAT, 50PBAT/B3, 50PLA/B3, and the 5050premixed/B3 samples.

Weight Loss %	Temperature (°C)		
	50PBAT/B3	50PLA/B3	5050Premixed/B3
1	279.89	272.53	287.10
5	334.46	331.74	338.73
10	346.19	345.63	348.89

Table 3.
Thermal degradation of 50PBAT/B3, 50PLA/B3 and the 5050premixed/B3 samples.

A similar trend was observed in the 5 and 10% weight loss temperatures as the lowest stability was observed for the 50PLA/B3, and the 5050premixed/B3 blend showed the highest temperature for the onset of degradation.

The degradation trend of the samples relates directly to the level of clay dispersion in the PLA phase. As it was explained earlier, the premixed samples showed the lowest clay dispersion, which shows the highest temperature for the onset of the degradation. The PLA phase of the 50PLA/B3 sample contained the largest content of dispersed organoclay and showed the lowest thermal stability.

Paul et al. [51] evaluated the hydrolytic degradation of PLA nanocomposites and observed the presence of organoclay accelerated the degradation of PLA. The highest degradation was observed in samples containing unmodified clay, which could be explained by the higher hydrophilicity of the particles. This could lead to an overall conclusion that clay's role in the acceleration of the degradation of PLA could be from the greater absorption of water by these particles [52, 53].

4. Conclusion and outlook

The increasing global demand for sustainable alternatives has led to a growing interest in biodegradable polymers as a solution to mitigate the environmental impact of traditional plastics. However, the field faces significant challenges, especially in enhancing the performance of biodegradable polymers to exceed conventional plastic benchmarks. Among these, poly(lactic acid) (PLA) stands out for its renewable sourcing but is hindered by brittleness, prompting study into blending with other biodegradable polymers such as poly(butylene adipate-co-terephthalate) (PBAT). Despite promising results in increased elongation, challenges persist, with blends demonstrating lower modulus and strength than pure PLA. The incorporation of organically modified nanoclay as a compatibilizing agent suggests a practical solution to overcome these limitations. However, the dispersion of nanoparticles within the blend, influenced by polymer–polymer interactions and mixing sequences, poses a significant challenge. Achieving a thermodynamic equilibrium state becomes difficult due to the high viscosity of polymers, affecting the location of organoclay particles and, consequently, the final properties of the blend. Understanding the complicated dynamics of mixing sequences is crucial for tailoring polymers with desired properties, making this an important area for further research in the field.

This work investigated the effect of sequential mixing of organically modified silicate layers on the rheology, morphology, and thermal properties of PBAT and PLA blends. The TEM results exhibited partial dispersion of the silicate layers in the blend. The premixed samples achieved the lowest level of clay dispersion among the samples studied. Better dispersion of the silicate layers in the 50PLA/B3 and 50PBAT/B3 indicates that a higher level of dispersion for clay can be achieved when it is mixed with a single polymer than when mixed with a blend of two polymers.

For blends with 50/50 ratio, solid-like behavior was observed in the melt state for the sequentially mixed composites but not for the premixed one. The rheological data are consistent with the quality of dispersion observed in the TEM images as the premixed blend showed the worst dispersion of the clay platelets and the lowest storage and loss modulus with highest dependence on the frequency.

Larger PLA domain sizes were detected in the 50PLA/B3 sample among the 50/50 ratio blends due to its partial continuity of the PLA phase dispersed in the PBAT matrix. Both phases were partially continuous in the 70PLA/B3 sample.

Conflict of interest

The authors declare no conflict of interest.

Author details

Mahin Shahlari[1,2], Bahareh Baheri[1] and Sunggyu Lee[1]*

1 Chemtech Innovators, LLC, Athens, OH, USA

2 Northern Technologies International Corporation, Circle Pines, MN, USA

*Address all correspondence to: drkblee@gmail.com

References

[1] Fiandra EF, Shaw L, Starck M, McGurk CJ, Mahon CS. Designing biodegradable alternatives to commodity polymers. Chemical Society Reviews. 2023;**52**(23):8085-8105

[2] Kumar R, Lalnundiki V, Shelare SD, Abhishek GJ, Sharma S, Sharma D, et al. An investigation of the environmental implications of bioplastics: Recent advancements on the development of environmentally friendly bioplastics solutions. Environmental Research. 2024;**1**(244):117707

[3] Sinha Ray S, Okamoto M. Polymer/layered silicate nanocomposites: A review from preparation to processing. Progress in Polymer Science. 2003;**28**(11):1539-1641

[4] Pluta M. Melt compounding of polylactide/organoclay: Structure and properties of nanocomposites. Journal of Polymer Science Part B: Polymer Physics. 2006;**44**(23):3392-3405

[5] Bhardwaj R, Mohanty AK. Modification of brittle polylactide by novel hyperbranched polymer-based nanostructures. Biomacromolecules. 2007;**8**(8):2476-2484

[6] Ko SW, Gupta RK, Bhattacharya SN, Choi HJ. Rheology and physical characteristics of synthetic biodegradable aliphatic polymer blends dispersed with MWNTs. Macromolecular Materials and Engineering. 2010;**295**(4):320-328

[7] Chen GX, Choi JB, Yoon JS. The role of functional group on the exfoliation of clay in poly(L-lactide). Macromolecular Rapid Communications. 2005;**26**(3):183-187

[8] Tripathi N, Misra M, Mohanty AK. Durable polylactic acid (PLA)-based sustainable engineered blends and biocomposites: Recent developments, challenges, and opportunities. ACS Engineering Au. 2021;**1**(1):7-38

[9] Coltelli MB, Bronco S, Chinea C. The effect of free radical reactions on structure and properties of poly(lactic acid) (PLA) based blends. Polymer Degradation and Stability. 2010;**95**(3):332-341

[10] Chen CC, Chueh JY, Tseng H, Huang HM, Lee SY. Preparation and characterization of biodegradable PLA polymeric blends. Biomaterials. 2003;**24**(7):1167-1173

[11] Chen BK, Shen CH, Chen SC, Chen AF. Ductile PLA modified with methacryloyloxyalkyl isocyanate improves mechanical properties. Polymer. 2010;**51**(21):4667-4672

[12] Hashima K, Nishitsuji S, Inoue T. Structure-properties of super-tough PLA alloy with excellent heat resistance. Polymer. 2010;**51**(17):3934-3939

[13] Coiai S, Di Lorenzo ML, Cinelli P, Righetti MC, Passaglia E. Binary green blends of poly(lactic acid) with poly(butylene adipate-co-butylene terephthalate) and poly(butylene succinate-co-butylene adipate) and their nanocomposites. Polymers. 2021;**13**(15):2489

[14] Jiang L, Wolcott MP, Zhang J. Study of biodegradable polylactide/poly(butylene adipate-co-terephthalate) blends. Biomacromolecules. 2006;**7**(1):199-207

[15] Signori F, Coltelli MB, Bronco S. Thermal degradation of poly(lactic acid) (PLA) and poly(butylene

adipate-co-terephthalate) (PBAT) and their blends upon melt processing. Polymer Degradation and Stability. 2009;**94**(1):74-82

[16] Lee S, Lee Y, Lee JW. Effect of ultrasound on the properties of biodegradable polymer blends of poly(lactic acid) with poly(butylene adipate-co-terephthalate). Macromolecular Research. 2007;**15**(1):44-50

[17] Paul DR, Robeson LM. Polymer nanotechnology: Nanocomposites. Polymer. 2008;**49**(15):3187-3204

[18] Ajayan PM, Schadler LS, Braun PV. Nanocomposite Science and Technology [Internet]. Hoboken, NJ: Wiley; 2003. Available from: https://books.google.com/books?id=lrZRAAAAMAAJ

[19] Baheri B, Lindenberger AL, Sharma S, Lee S. Characterization of linear low-density polyethylene and halloysite nanotube (LLDPE/HNT) composites based on two-roll calendering melt fabrication. Journal of Applied Polymer Science. 2023;**140**(5):e53259

[20] Krishnamachari P, Zhang J, Lou J, Yan J, Uitenham L. Biodegradable poly(lactic acid)/clay nanocomposites by melt intercalation: A study of morphological, thermal, and mechanical properties. International Journal of Polymer Analysis and Characterization. 2009;**14**(4):336-350

[21] Jiang L, Liu B, Zhang J. Properties of poly(lactic acid)/poly(butylene adipate-co-terephthalate)/nanoparticle ternary composites. Industrial and Engineering Chemistry Research. 2009;**48**(16):7594-7602

[22] Ko SW, Hong MK, Park BJ, Gupta RK, Choi HJ, Bhattacharya SN. Morphological and rheological characterization of multi-walled carbon nanotube/PLA/PBAT blend nanocomposites. Polymer Bulletin. 2009;**63**(1):125-134

[23] da Costa FAT, Parra DF, Cardoso ECL, Güven O. PLA, PBAT, cellulose nanocrystals (CNCs), and their blends: Biodegradation, compatibilization, and nanoparticle interactions. Journal of Polymers and the Environment. 2023;**31**(11): 4662-4690

[24] He H, Liu B, Xue B, Zhang H. Study on structure and properties of biodegradable PLA/PBAT/organic-modified MMT nanocomposites. Journal of Thermoplastic Composite Materials. 2022;**35**(4):503-520

[25] Sinha Ray S, Maiti P, Okamoto M, Yamada K, Ueda K. New polylactide/layered silicate nanocomposites. 1. Preparation, characterization, and properties. Macromolecules. 2002;**35**(8):3104-3110

[26] Sinha Ray S, Yamada K, Okamoto M, Ogami A, Ueda K. New polylactide/layered silicate nanocomposites. 3. High-performance biodegradable materials. Chemistry of Materials. 2003;**15**(7):1456-1465

[27] Chang JH, An YU, Sur GS. Poly(lactic acid) nanocomposites with various organoclays. I. Thermomechanical properties, morphology, and gas permeability. Journal of Polymer Science Part B: Polymer Physics. 2003;**41**(1):94-103

[28] Alves JL, PTV D e R, Realinho V, Antunes M, Velasco JI, Morales AR. Influence of chemical composition of Brazilian organoclays on the morphological, structural and thermal properties of PLA-organoclay nanocomposites. Applied Clay Science. 2019;**1**(180):105186

[29] Alves JL, PTV D e R, Realinho V, Antunes M, Velasco JI, Morales AR. The effect of Brazilian organic-modified montmorillonites on the thermal stability and fire performance of organoclay-filled PLA nanocomposites. Applied Clay Science. 2020;**1**(194):105697

[30] Fenouillot F, Cassagnau P, Majesté JC. Uneven distribution of nanoparticles in immiscible fluids: Morphology development in polymer blends. Polymer. 2009;**50**(6):1333-1350

[31] Liebscher M, Domurath J, Saphiannikova M, Müller MT, Heinrich G, Pötschke P. Dispersion of graphite nanoplates in melt mixed PC/SAN polymer blends and its influence on rheological and electrical properties. Polymer. 2020;**18**(200):122577

[32] Kim SW, Jo WH, Lee MS, Ko MB, Jho JY. Preparation of clay-dispersed poly(styrene-co-acrylonitrile) nanocomposites using poly(ϵ-caprolactone) as a compatibilizer. Polymer. 2001;**42**(24):9837-9842

[33] Ramírez-Vargas E, Valera-Zaragoza M, Sánchez-Valdes S, Hernández-Valdez JS, Ibarra-Castillo FF. Effect of processing conditions on the structural morphology of PP–EP/EVA/organoclay ternary nanocomposites. Polymer Bulletin. 2009;**62**(3):391-403

[34] Dasari A, Yu ZZ, Mai YW. Effect of blending sequence on microstructure of ternary nanocomposites. Polymer. 2005;**46**(16):5986-5991

[35] Tao JR, Yang D, Yang Y, He QM, Fei B, Wang M. Migration mechanism of carbon nanotubes and matching viscosity-dependent morphology in co-continuous poly(lactic acid)/poly(ε-caprolactone) blend: Towards electromagnetic shielding enhancement. Polymer. 2022;**14**(252):124963

[36] Vo LT, Giannelis EP. Compatibilizing poly(vinylidene fluoride)/nylon-6 blends with nanoclay. Macromolecules. 2007;**40**(23):8271-8276

[37] Ha MH, Kim MS, Kim BK, Kim W, Lee MC, Kim HD. Effects of the blending sequence in polyolefin ternary blends. Journal of Applied Polymer Science. 2004;**92**(2):804-811

[38] Wang K, Wang C, Li J, Su J, Zhang Q, Du R, et al. Effects of clay on phase morphology and mechanical properties in polyamide 6/EPDM-g-MA/organoclay ternary nanocomposites. Polymer. 2007;**48**(7):2144-2154

[39] Torres RA, Arellano-Ceja J, Hernández-Hernández ME, González-Núñez R. Effects of the blending sequence and interfacial agent on the morphology and mechanical properties of injection molded PC/PP blends. Polymer Bulletin. 2007;**59**(2):251-260

[40] Yoo Y, Tiwari RR, Yoo YT, Paul DR. Effect of organoclay structure and mixing protocol on the toughening of amorphous polyamide/elastomer blends. Polymer. 2010;**51**(21):4907-4915

[41] Someya Y, Kondo N, Shibata M. Biodegradation of poly(butylene adipate-co-butylene terephthalate)/layered-silicate nanocomposites. Journal of Applied Polymer Science. 2007;**106**(2):730-736

[42] Shahlari M. Biodegradable Polymer/Clay Nanocomposites Based on Poly(Butylene Adipate-co-Terephthalate) and Polylactide [Thesis] [Internet]. USA: Missouri University of Science and Technology; 2011. Available from: https://scholarsmine.mst.edu/doctoral_dissertations/11

[43] Utracki LA. Clay-Containing Polymeric Nanocomposites. Vol. 1. Shrewsbury, Shropshire, United

Kingdom: iSmithers Rapra Publishing; 2004

[44] Vermant J, Ceccia S, Dolgovskij MK, Maffettone PL, Macosko CW. Quantifying dispersion of layered nanocomposites via melt rheology. Journal of Rheology. 2007;**51**(3):429-450

[45] Zhao J, Morgan AB, Harris JD. Rheological characterization of polystyrene–clay nanocomposites to compare the degree of exfoliation and dispersion. Polymer. 2005;**46**(20):8641-8660

[46] Krishnamoorti R, Giannelis EP. Rheology of end-tethered polymer layered silicate nanocomposites. Macromolecules. 1997;**30**(14):4097-4102

[47] Elias L, Fenouillot F, Majesté JC, Martin G, Cassagnau P. Migration of nanosilica particles in polymer blends. Journal of Polymer Science Part B: Polymer Physics. 2008;**46**(18):1976-1983

[48] Filippone G, Dintcheva NTz, La Mantia FP, Acierno D. Using organoclay to promote morphology refinement and co-continuity in high-density polyethylene/polyamide 6 blends – Effect of filler content and polymer matrix composition. Polymer. 2010;**51**(17):3956-3965

[49] Kontopoulou M, Liu Y, Austin JR, Parent JS. The dynamics of montmorillonite clay dispersion and morphology development in immiscible ethylene–propylene rubber/polypropylene blends. Polymer. 2007;**48**(15):4520-4528

[50] Sinha Ray S, Bandyopadhyay J, Bousmina M. Effect of organoclay on the morphology and properties of poly(propylene)/poly[(butylene succinate)-co-adipate] blends. Macromolecular Materials and Engineering. 2007;**292**(6):729-747

[51] Paul MA, Delcourt C, Alexandre M, Degée P, Monteverde F, Dubois P. Polylactide/montmorillonite nanocomposites: Study of the hydrolytic degradation. Polymer Degradation and Stability. 2005;**87**(3):535-542

[52] Ray SS, Yamada K, Okamoto M, Ueda K. Control of biodegradability of polylactide via nanocomposite technology. Macromolecular Materials and Engineering. 2003;**288**(3):203-208

[53] Zhou Q, Xanthos M. Nanoclay and crystallinity effects on the hydrolytic degradation of polylactides. Polymer Degradation and Stability. 2008;**93**(8):1450-1459

Section 3

Environmental and Industrial Applications of Clays and Polymer–Clay Nanocomposites

Chapter 4

Removal Efficiency of Phenolic Compounds (Bisphenol A and Pentachlorophenol) by Adsorption Using a Bentonite-CTAB

Nouhaila Hadoudi, M'hamed Ahari, Najlae Zaki, Asmae Charki, Hossain El Ouarghi, Ayoub Bayoussef, Mohammed Mansori, Mourabit Fouad, Amin Salhi and Hassan Amhamdi

Abstract

Bentonite clay modified with cetyltrimethylammonium bromide (CTAB) has been investigated as an effective adsorbent for the removal of bisphenol A (BPA) and pentachlorophenol (PCP) from aqueous solutions. The adsorption performance of Bentonite-CTAB was evaluated by conducting batch adsorption experiments under different conditions. The adsorption isotherms of BPA and PCP on Bentonite-CTAB (BT-CTAB) were analyzed using the Langmuir and Freundlich models. The Langmuir model provided a better fit to the experimental data, suggesting the presence of monolayer adsorption. The adsorption kinetics of BPA and PCP on Bentonite-CTAB were studied using pseudo-first-order and pseudo-second-order models. The results indicate that the adsorption process follows pseudo-second-order kinetics, suggesting that the adsorption is controlled principally by chemisorption. Equilibrium time for both pollutants was achieved within 30–40 min. The results of adsorption studies indicated that Bentonite-CTAB exhibited excellent adsorption capacity for bisphenol A and pentachlorophenol. The high surface area and presence of active sites on Bentonite-CTAB favored adsorption of pollutants from aqueous solution. The adsorption process adopted pseudo-second-order kinetics, indicating the involvement of chemisorption. the adsorption isotherms of BPA and PCP on Bentonite-CTAB were analyzed using the Langmuir and Freundlich models. The Langmuir model provided a better fit to the data, suggesting monolayer adsorption.

Keywords: bisphenol A, pentachlorophenol, bentonite, BT-CTAB, adsorption

1. Introduction

Adsorption, a technique commonly used to remove pollutants from contaminated water sources, including phenolic compounds, has gained in popularity. Bentonite, a clay mineral known for its high surface area and adsorption capacity, has been widely studied for its potential in adsorption processes. However, the pure form of Bentonite may have limitations in effectively adsorbing phenolic compounds, leading researchers to explore modifications to its surface properties using surfactants. Cetyltrimethylammonium bromide (CTAB), a cationic surfactant commonly used to modify clay minerals, showed promise in improving the adsorption performance of Bentonite. By enhancing hydrophobic interactions and increasing surface area, the addition of CTAB has improved the adsorption of phenolic compounds.

Recent studies have focused on the use of the Bentonite-CTAB composite for the adsorption of phenolic compounds. For example, Khenifi et al (2009) investigated removal of 2,4-dichlorophenol (2,4-DCP) from aqueous solutions by Bentonite CTAB and observed a significantly higher adsorption capacity. The improved performance was attributed to enhanced hydrophobic interactions and increased accessibility of phenolic compounds to the surface of the modified clay [1]. Similarly, Noufel et al. (2020) examined the adsorption of bisphenol A with a Bentonite-CTAB composite, reporting a substantial improvement in removal efficiency over pure Bentonite [2].

The use of modified Moroccan bentonite is highly effective in removing pollutants and improving water quality [3, 4], Bentonite used as a coagulant has a basic pH probably due to the presence of potassium feldspar and calco-sodium. Coagulant/bentonite mixtures in surface water clarification have very important consequences [5–7].

This study aims to evaluate the removal efficiency of phenolic compounds such as bisphenol A and pentachlorophenol using a Bentonite-CTAB composite. By manipulating experimental conditions and conducting batch adsorption experiments, various factors influencing the adsorption process, as well as adsorption capacity and kinetics, will be evaluated. Isothermal and kinetic models will be used to better understand the adsorption mechanism and identify the steps that control the rate. This study will provide valuable information on the performance and optimization of the Bentonite-CTAB composite for the removal of phenolic compounds.

2. Methodology

2.1 Bentonite-CTAB preparation

The synthesis procedure for the Bentonite-CTAB modification:

- Activation of Bentonite: Add 50 g of bentonite sample and 12% hydrochloric acid solution (HCl) to the beaker to completely cover the bentonite and stir the mixture at 60°C for 4 h to activate. Centrifuge the mixture to separate the activated bentonite from the supernatant, discard and save the supernatant. Neutralize the activated bentonite, add 10% sodium hydroxide solution (2.5 mol/L) (NaOH) to neutralize, stir the mixture and repeat the neutralization step until the bentonite is neutral. The neutralized bentonite is washed 4–5 times alternately with deionized water and neutralized with ethanol. dry the final sample at 80°C and store it for later use.

- Preparation of CTAB solution: 10 g of CTAB add it to a volume of deionized water to dissolve the CTAB while stirring, continue stirring until a homogeneous solution is obtained.

- Modification of Bentonite with CTAB: Take 20 g of the activated Bentonite, add 700 mL of distilled water and stir the mixture at 25°C for 4 h to completely hydrated the bentonite, Defined volume (500 mL) of 6 g/L CTAB solution was slowly added to the above mixture at 60°C continue stirring for 4 h, wash the sample with distilled water three times to remove residual CTAB, repeat the washing step until a 0.1 mol/L silver nitrate solution,used to detect precipitation, shows no precipitation in the washing solution, dry the final modified sample at 80°C, store the modified Bentonite-CTAB for later use.

2.2 Bisphenol A (BPA) and Pentachlorophenol (PCP)

See **Table 1**

2.3 Determination of contaminant concentrations in waste water

A 10 ml sample of the pollutant was placed in a quartz cuvette and analyzed using a UV-VIS spectrophotometer. To determine the concentrations of bisphenol A (BPA) and pentachlorophenol (PCP) after each experiment, calibration curves for BPA and PCP were first set up. Six different concentrations of pure BPA and PCP were prepared, and their absorbance values were measured using a UV-Vis spectrometer at their respective wavelengths of maximum absorption (λ max = 276 nm for BPA and 320 nm for PCP). The maximum absorbance values obtained were plotted against each concentration of BPA and PCP. Based on this plot, the concentrations of BPA and PCP present in the wastewater sample were calculated.

2.4 Batch adsorption study

Batch experiments were conducted to assess the adsorption efficiency of Bentonite-CTAB for the removal of Bisphenol A (BPA) and Pentachlorophenol (PCP). A BPA and PCP solution was prepared by dissolving a specific quantity of solid in deionized water at a concentration of 1 g/L.

Proprieties	BPA	PCP
Chemical formula	$C_{15}H_{16}O_2$	C_6HCl_5O
Molar mass	228,28 g/mol	266.35 g/mol
Water solubility	120–300 mg/l at pH 7	0.020 g/L at 30°C
Melting point	158°C	190°C
Boiling point	250–252°C	310°C
Vapor pressure	5×10^{-6} Pa (25°C)	0.0001 mmHg (25°C)

Table 1.
Properties of BPA and PCP.

Adsorption experiments were performed by adding a fixed amount of 0.1 g of Bentonite-CTAB adsorbent to a series of 100 ml Erlenmeyer flasks containing 50 ml of the BPA and PCP solutions. The flasks were then stirred until equilibrium was attained. At specified time intervals, 10 ml samples were collected and filtered to eliminate any residual adsorbent present in the solution. The concentrations of the liquid samples were subsequently analyzed using a UV spectrophotometer (Rayleigh, UV-1800) at the maximum wavelength of BPA and PCP.

Several parameters were considered to study their influence on the adsorption capacity. These parameters included contact time (ranging from 10 to 120 min), initial concentration of the BPA and PCP solution (ranging from 10 to 100 ppm), pH values (ranging from 3 to 11), and temperature (ranging from 25 to 65°C).

To determine the adsorption capacity of Bentonite-CTAB for BPA and PCP, batch adsorption experiments were conducted. The concentration of BPA and PCP were determined using ultraviolet-visible spectrophotometry. The adsorption capacity (q_e) of BPA and PCP on Bentonite-CTAB were calculated using equations (1):

$$q_e = V \times \frac{\left(C_i - C_f\right)}{w} \tag{1}$$

Eq. (1) determines the equilibrium adsorption capacity (q_e) of Bentonite-CTAB for the BPA and PCP solution based on the volume (V), initial and final concentrations (C_i and C_f), and the mass of the adsorbent (m).

These equations allow for the quantification of the adsorption performance of Bentonite-CTAB for BPA and PCP and facilitate the assessment of different experimental conditions on the adsorption process.

3. Discussion

3.1 Adsorption studies of Bentonite-CTAB

3.1.1 Adsorption kinetics experiments

Adsorption kinetics is an important parameter for understanding the mechanism of BPA and PCP adsorption onto the adsorbent surface. Batch adsorption versus time data using a fixed dose of adsorbent 0.1 g at 25°C were used for kinetic modeling of BPA and PCP with an initial concentration of 1000 mg/L.

Therefore, 30 min for BPA and 40 min for PCP, a contact time sufficient to reach adsorption equilibrium, was chosen for further adsorption studies. To study the adsorption kinetics, pseudo-first-order and pseudo-second-order models were applied to the experimental data in **Figure 1**. Correlation coefficients and adsorption kinetic parameters are summarized in **Table 2**. These results indicated that the adsorption of BPA and PCP on BT-CTAB can be explained by pseudo-second-order kinetics.

Contact time is an important parameter in the adsorption process. As shown in **Figure 2**, the adsorption of BPA and PCP on Bentonite-CTAB changes with increasing contact time. The results show that the rate of adsorption is initially high and then gradually slows down until equilibrium is reached. The adsorption capacities of

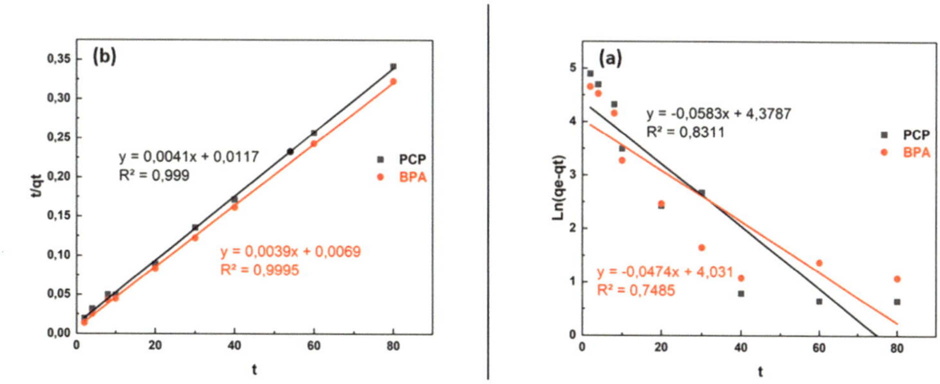

Figure 1.
Pseudo first order (a), Pseudo second order (b).

Model	Equation	Value of parameters			
		BPA		PCP	
Pseudo-first-order	$\ln(q_e - q_t) = \ln q_e - k_1 * t$	$q_{e,cal}$ (mg/g) k_1 (min^{-1}) R^2	56.36 0.0002 0.74	$q_{e,\,cal}$ (mg/g) k_1 (min^{-1}) R2	79.73 0.0003 0.83
Pseudo-second-order	$\frac{t}{q_e} = \frac{1}{k_2 \times q_e^2} + \frac{1}{q_e} * t$	$q_{e,cal}$ (mg/g) k_2 (g/mg/min) R^2	256.41 0.0022 0.99	$q_{e,\,cal}$ (mg/g) k_2(g/mg/min) R^2	243.90 0.0014 0.99

Table 2.
Kinetic model parameters for the adsorption process.

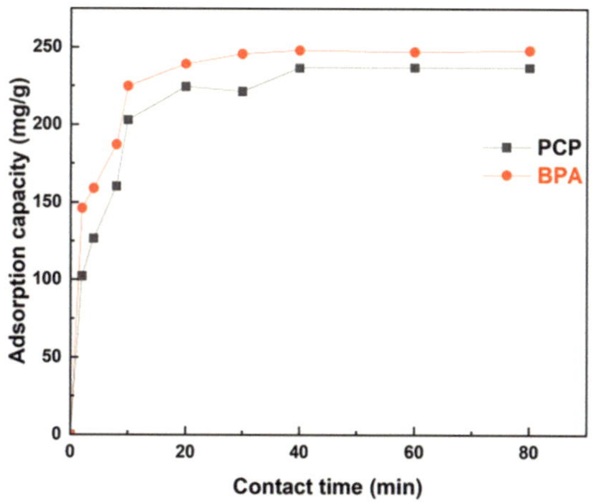

Figure 2.
Influence of contact time on adsorption of BPA and PCP.

BPA and PCP are very high, about 248 mg/g and 233 mg/g respectively. Adsorption equilibrium is reached in 30 min for BPA and 40 min for PCP, a study conducted by Ahari et al. in 2020 demonstrated that equilibrium sorption of chlorophenols onto geosynthetic clay liners' geotextile was achieved after 72 h [8].

Table 2 shows the experimental and calculated parameters for the pseudo-first-order and pseudo-second-order equations. The R^2 values of the pseudo-first-order model showed no consistent trend, and the experimental q_e values ($q_{e, exp}$) did not agree well with the calculated values ($q_{e, cal}$) obtained from the linear plots. This indicates that the resulting adsorption of the two contaminants on all adsorbents does not follow a pseudo-first-order kinetic model. However, all R^2 values obtained from the pseudo-second-order model are close to 1, and the experimental q-values and the calculated values obtained from the linear plots are in good agreement. These results suggest that the second pseudo-order model better describes the experimental data. According to previous studies, the second pseudo-order model assumes that the rate-limiting step is the chemical bonding process in which electrons are shared or exchanged between adsorbent and adsorbate.

3.1.2 Adsorption isotherm experiments

Experimental equilibrium adsorption data for BPA and PCP on Bentonite-CTAB were modeled using Langmuir and Freundlich isotherms. A comparison of the experimental BPA and PCP adsorption data with the theoretical Langmuir and Freundlich isotherm plots is shown in **Figure 3**. **Table 3** shows the optimal parameters for each model.

The experimental adsorption isotherms were compared with the theoretical models of Langmuir and Freundlich. Batch experiments were performed using 0.1 g of the adsorbent with different initial concentrations of PCP (10–1000 mg/L) under continuous stirring for a contact time of 40 min and at pH = 7. The Langmuir isotherm model is one of the most common adsorption models. It assumes monolayer coverage of the adsorbate on a homogeneous adsorbent surface; at equilibrium, a saturation point is reached where no further adsorption can occur. The essential characteristics and feasibility of the Langmuir isotherm are described in terms of a dimensionless constant separation factor (R_L), which is defined as follows:

$$R_L = \frac{1}{1 + C_e * K_L} \quad (2)$$

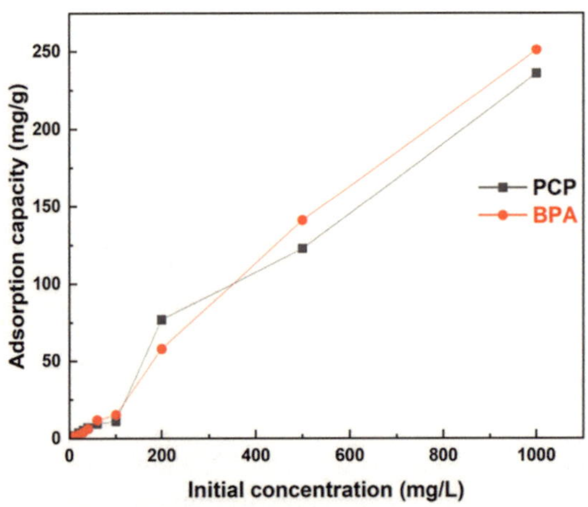

Figure 3.
Influence of initial concentration on the adsorption of BPA and PCP.

Isotherm	Equation	Value of parameters			
		BPA		PCP	
Langmuir	$\frac{1}{q_e} = \frac{1}{k_L q_{\max}} * \frac{1}{C_e} + \frac{1}{q_{\max}}$	q_m(mg/g)	15.22	q_m(mg/g)	36.63
		K_L(L/mg)	0.016	K_L(L/mg)	0.029
		R_L	0.066	R_L	0.032
		R^2	0.91	R^2	0.80
Freundlich	$\log q_e = \log k_F + \frac{1}{n} \log C_e$	K_F (mg/g)	0.014	K_F (mg/g)	18.32
		n	0.247	n	0.879
		R^2	0.85	R^2	0.39

Table 3.
Langmuir and Freundlich constants for the adsorption of BPA and PCP onto BT-CTAB at temperature of 25°.

The Langmuir model was in better agreement with the experimental data than the Freundlich model for both contaminants, with R^2 = 0.91 for BPA and R^2 = 0.80 for PCP. Moreover, the R_L is greater than zero, indicating good adsorption for both pollutants studied. This result means that pollutants can be homogeneously adsorbed by the active receptor sites on the surface of the Bentonite-CTAB adsorbent and a saturated adsorption capacity is expected due to the limited number of adsorbent sites on the surface [9]. This indicates that monolayer adsorption of the BPA and PCP occurred on surfaces that were uniform in terms of adsorption affinity. In this study, the maximum adsorption capacities for BPA and PCP resulting from the initial concentration effects were found to be 251.05 mg/g and 235.85 mg/g, respectively. Djebri et al. (2017) also studied the adsorption of an endocrine disruptor onto modified Bentonite clay. The authors identified that the Langmuir isotherm adequately fitted the adsorption data and that the maximum adsorption capacity was 127.7 mg/g for bisphenol A (**Figure 4**) [10].

3.1.3 Effect of temperature and thermodynamics studies

As shown in **Figure 5** The adsorption capacity of BPA and PCP decreases with increasing temperature in the range of 25–65°C. Thus, the adsorption capacity of BPA and PCP is high at low temperatures. However, at higher temperatures, a decrease in the adsorption capacity of BPA and PCP is observed. This behavior, in agreement

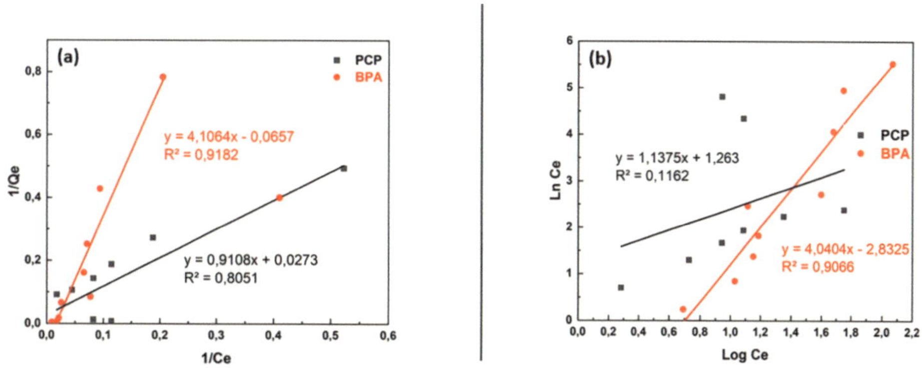

Figure 4.
Langmuir Isotherm (a); Freundlich Isotherm (b).

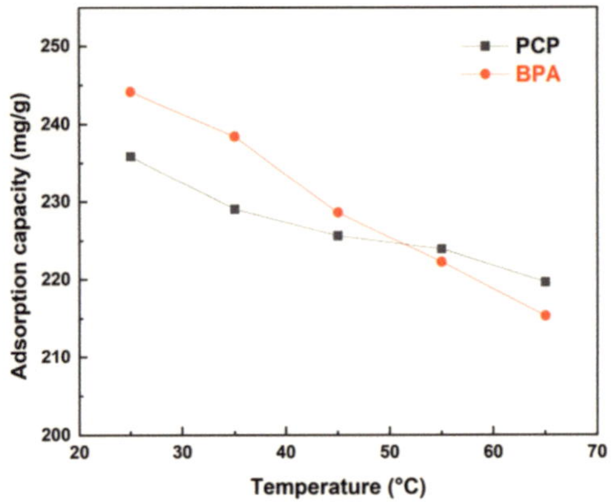

Figure 5.
Influence of temperature on the adsorption of BPA and PCP.

System	$\Delta H°$ (kJ/mol/K)	$\Delta S°$ (lJ/mol/K)	$\Delta G°$ (kJ/mol)				
			298	308	318	328	338
BPA	−40.14	−116.36	−5.8	−4.19	−2.59	−1.89	−1.23
PCP	−16.16	−43.23	−3.54	−2.57	−2.22	−2.08	−1.66

Table 4.
Thermodynamic parameters for adsorption of BPA and PCP.

with Arrhenius' law, suggests that the BPA and PCP adsorption reaction, occurring on the Bentonite-CTAB surface, is exothermic, and indicates that with each increase in temperature.

Table 4 shows the calculated values of the adsorption amounts $\boldsymbol{\Delta G°}$, $\boldsymbol{\Delta H°}$, and $\boldsymbol{\Delta S°}$ of BPA and PCP on bentonite CTAB. A negative value of $\boldsymbol{\Delta H°}$ indicates that the adsorption process is exothermic in nature. In general, the enthalpy of adsorption for physisorption is typically between 0 and −42 kJ/mol, whereas for chemisorption it is between −42 and −125 kJ/mol [11]. Therefore, the adsorption of BPA and PCP in this study can also be considered as physisorption. Negative $\boldsymbol{\Delta S°}$ values indicate the affinity of Bentonite-CTAB for BPA and PCP in aqueous solutions and may indicate structural changes of the adsorbates and adsorbents during the adsorption process [11]. The $\boldsymbol{\Delta G°}$ values for BPA and PCP were negative, suggesting that the process was spontaneous in nature and that the spontaneity decreased as the temperature increased from 25°C to 65°C (**Figure 6**).

3.1.4 Effect of solution pH

The initial pH of the solution plays a special role on adsorption because of its effect on the solution chemistry and the degree of ionization of the adsorbent surface. BPA and PCP removal was studied by varying the pH of the aqueous solution from 3 to 11 at 25°C with a fixed BPA and PCP concentration of 1000 mg/L and an adsorbent mass of 0.1 mg. The **Figure 7** shows the adsorption capacity of BPA and PCP by Bentonite-CTAB at different pH values. As can be seen from the graph, the removal of BPA and PCP increased significantly as the pH of the starting solution was increased from 1 to

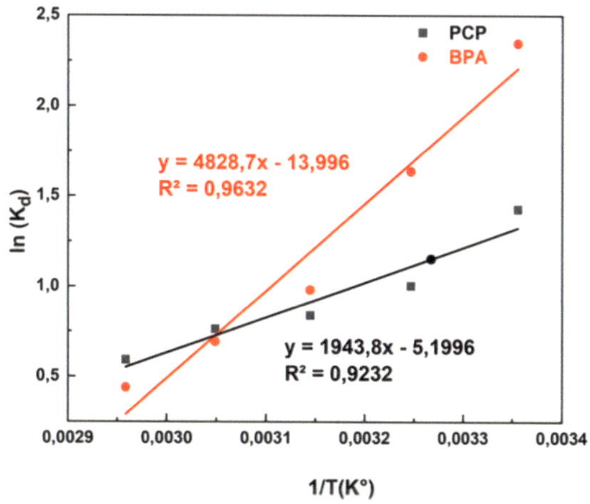

Figure 6.
Thermodynamic fitting curve of the adsorbed BPA and PCP.

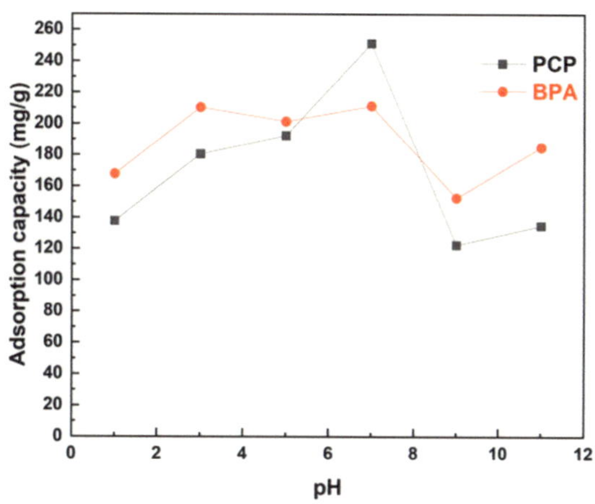

Figure 7.
Influence of pH on the adsorption of BPA and PCP.

7, and then decreased from 7 to 11 as the pH increased. In this study, maximal removal of BPA and PCP was achieved at pH 7, with maximal removal of 200 mg/g for BPA and 250 mg/g for PCP.

The strong adsorption of BPA and PCP observed at pH (1–7) is due in part to the electrostatic interaction between several positively charged surface functional groups and negatively charged chlorophenolate ions for the process. BPA and PCP adsorption. As the pH moved into the more basic range, The adsorption capacity of bisphenol A (BPA) and pentachlorophenol (PCP) decreases significantly at pH values above 7. This phenomenon is due to the simultaneous decrease of negative surface charge and carbon content with increasing pH. As ionization increases and the surface becomes negatively charged, the electrostatic affinity between the adsorbent and the pollutant weakens. As a result, the electrostatic interaction weakens and the adsorption efficiency decreases. Adjustment of pH value has been proven to be a strategic regulator that influences the special physicochemical properties of pollutants and adsorbents [12].

3.1.5 Mechanism for BPA and PCP on Bentonite-CTAB

The adsorption mechanism for BPA and PCP on Bentonite-CTAB involved multiple interactions, including electrostatic attraction, and hydrogen bonding. The CTAB cations on the surface of Bentonite-CTAB provided positive charges that interacted with the negatively charged pollutants, facilitating their adsorption. Additionally, the presence of hydroxyl groups in the pollutant molecules enabled hydrogen bonding with the functional groups on the Bentonite-CTAB surface.

Initially, the active sites on the Bentonite-CTAB surface facilitate adsorption, with hydrophobic interactions likely playing a role in the initial binding of organic components from BPA and PCP. Electrostatic forces come into play as positively charged CTAB ions on the Bentonite surface interact with negatively charged functional groups on the pollutants, promoting electrostatic adsorption. Additionally, hydrogen bond formation between specific functional groups in BPA and PCP and active sites on Bentonite-CTAB enhances the adsorption process. Molecular rearrangement may occur as contaminants adjust to the surface, optimizing adsorption efficiency. Ultimately, as the active sites become saturated, adsorption reaches an equilibrium point, signaling a balance between the fixation of contaminants and their release. It's important to recognize that this proposed mechanism simplifies a complex process, and experimental validation would be crucial to refining and confirming these interactions (**Figure 8**).

3.1.6 Comparison of adsorption capacity

Adsorption capacity and removal efficiency of contaminants in water Bisphenol A (BPA) and pentachlorophenol (PCP) are important factors in evaluating the effectiveness of various water treatment methods. BPA is an endocrine-disrupting compound commonly found in spring water, and PCP is a toxic chlorine compound commonly resulting from industrial activities. Several studies have investigated the adsorption performance of different materials for the removal of BPA and PCP. Adsorption processes involve the attraction of contaminants to solid surfaces. Studies show that the adsorption capacity of BPA is often higher than that of PCP due to their different molecular properties and interactions with adsorbents. However, specific results depend on factors such as sorbent type, pH, and initial concentration.

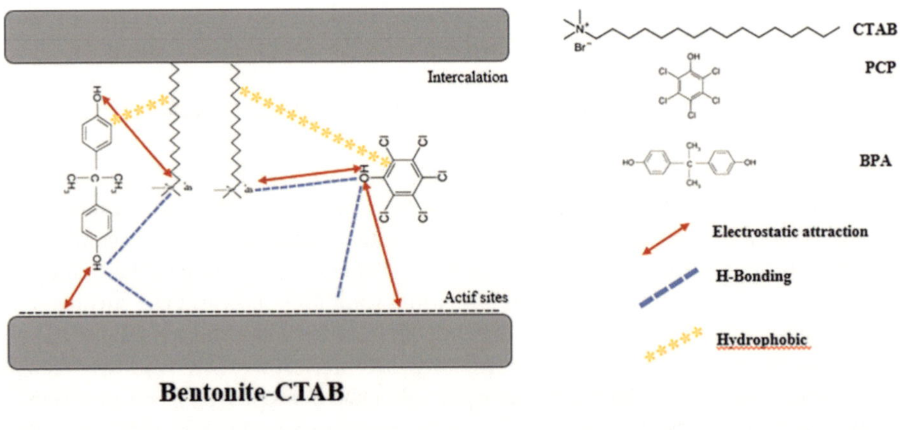

Figure 8.
Adsorption mechanism of Bentonite-CTAB on the removal of BPA and PCP.

Adsorbate	Adsorbent	q_{max} (mg/g)	Reference
PCP	Bentonite-CTAB	243.90	This study
	Bentonite	10,63	[13]
	Magnetic recoverable carbon	120.48	[14]
	Carbon nanotube hybrid polymer adsorbents	20–100	[15]
	22-2-10 modified vermiculite	85	[16]
	modified chitosan	24.37	[17]
BPA	Bentonite-CTAB	256,41	This study
	Organobentonite	127.7	[10]
	Calcium alginate/organo activated bentonite (A-OAB)	252.9	[2]
	Fe3O4-polyaniline	28.1	[18]
	sulfur-doped titanium dioxide on magnetic bentonite (CST/γ-Fe2O3-BT)	77.36	[19]
	Synthesized Chitosan (SC)	34.48	[20]

Table 5.
Comparison of adsorption capacity.

Table 5 presents the results of various studies comparing the adsorption capacities of his BPA and PCP using different adsorbent materials. The results show that BPA generally has a higher adsorption capacity than PCP under similar conditions, although specific values may vary depending on the sorbent used and initial contaminant concentration.

4. Conclusion

The study investigated the use of Bentonite clay modified with cetyltrimethylammonium bromide (CTAB) as an effective adsorbent for the removal of bisphenol A (BPA) and pentachlorophenol (PCP) from aqueous solutions. The modified clay, Bentonite-CTAB, demonstrated high adsorption potential for both pollutants, the adsorption isotherms of BPA and PCP on Bentonite-CTAB were analyzed using the Langmuir and Freundlich models. The Langmuir model provided a better fit to the data, suggesting monolayer adsorption. The maximum adsorption capacities were determined to be 243,90 mg/g for PCP and 256,41 mg/g for BPA, indicating the significant adsorption capabilities of Bentonite-CTAB for these contaminants.

The adsorption kinetics of BPA and PCP on Bentonite-CTAB were studied using pseudo-first-order and pseudo-second-order models. The results indicated that the adsorption process followed pseudo-second-order kinetics, indicating chemisorption as the principal mechanism. Equilibrium was reached within 30 min for BPA and 40 min for PCP.

The study demonstrated that Bentonite-CTAB exhibited excellent adsorption capacity for BPA and PCP due to its high surface area and presence of active sites. The adsorption process was found to be pH-dependent, with higher adsorption capacity observed at lower pH values, attributed to increased positive charge on the surface of Bentonite-CTAB, facilitating electrostatic interactions with the pollutants. Temperature also influenced the adsorption process, with lower temperatures favoring higher adsorption capacities.

The adsorption mechanism for BPA and PCP on Bentonite-CTAB involved multiple interactions, including electrostatic attraction and hydrogen bonding. The CTAB cations on the surface of Bentonite-CTAB provided positive charges that interacted with the negatively charged pollutants, facilitating their adsorption. Additionally, the presence of hydroxyl groups in the pollutant molecules enabled hydrogen bonding with the functional groups on the Bentonite-CTAB surface.

Overall, the study highlights the effectiveness of Bentonite-CTAB as an adsorbent for the removal of BPA and PCP from aqueous solutions. The findings contribute to the understanding of the adsorption process and provide valuable insights for the development of efficient water treatment methods targeting these pollutants.

Author details

Nouhaila Hadoudi[1*], M'hamed Ahari[1*], Najlae Zaki[1], Asmae Charki[2], Hossain El Ouarghi[2], Ayoub Bayoussef[3,4], Mohammed Mansori[3,5], Mourabit Fouad[1], Amin Salhi[1] and Hassan Amhamdi[1]

1 Applied Chemistry Unit, Department of Chemistry, Faculty of Sciences and Techniques, Abdelmalek Essaâdi University, Al Hoceima, Morocco

2 Water and Environment Management Unit, National School of Applied Sciences, Abdelmalek Essaâdi University, Al Hoceima, Morocco

3 Faculty of Sciences and Technologies, Cadi Ayyad University, Marrakech, Morocco

4 Faculty of Sciences and Technology, Hassan First University of Settat, Settat, Morocco

5 IMED, Laboratory of Innovative Materials, Energy, and Sustainable Development, Faculty of Science and Technology, Cadi Ayyad University, Marrakech, Morocco

*Address all correspondence to: nouhailahadoudi79@gmail.com and m.ahari@uae.ac.ma

References

[1] Khenifi A, Zohra B, Kahina B, Houari H, Zoubir D. Removal of 2, 4-DCP from wastewater by CTAB/bentonite using one-step and two-step methods: A comparative study. Chemical Engineering Journal. 2009;**146**(3):345-354. DOI: 10.1016/j.cej.2008.06.028

[2] Noufel K, Djebri N, Boukhalfa N, Boutahala M, Dakhouche A. Removal of bisphenol A and trichlorophenol from aqueous solutions by adsorption with organically modified Bentonite, activated carbon composites: A comparative study in single and binary systems. Groundwater for Sustainable Development. 2020;**11**:100477. DOI: 10.1016/j.gsd.2020.100477

[3] Hadoudi N, Charki A, El Ouarghi H, Mourabit F, Salhi A, Amhamdi H, et al. Sorption of Bisphenol A from aqueous solutions by acid-activated bentonite clay. Desalination and Water Treatment. 2023;**285**:121-128. DOI: 10.5004/dwt.2023.29295

[4] Ahari M, Ddahim H, Ramadane R. Performance of bentonite clay as a coagulation aid on water quality. Desalination and Water Treatment. 2019;**143**:229-234. DOI: 10.5004/dwt.2019.23552

[5] Zaki N, Hadoudi N, Charki A, Bensitel N, Ouarghi H. E, Amhamdi H, Ahari M. H. Advancements in the chemical treatment of potable water and industrial wastewater using the coagulation–flocculation process. Separation Science and Technology. 2023;**58**:1-12. DOI: 10.1080/01496395.2023.2219381

[6] Charki A, El Ouarghi H, Ahari M. Synthesis of Leachate from the Al Hoceima Controlled Landfill and Characterization (Morocco, North of Africa). Moroccan. Journal of Chemistry. 2022;**10**(4):800-807. DOI: 10.48317/IMIST.PRSM/morjchem-v10i3.30231

[7] Charki A, El Ouarghi H, Ahari M. Treatability tests of synthetic leachate from the Al-Hoceima City Controlled Landfill. E3S Web of Conferences. EDP Sciences. 2021:07007. DOI: 10.1051/e3sconf/202131407007

[8] Ahari MH, Touze-Foltz N, Mazéas L. Sorption of Chlorophenols on Geotextile of the Geosynthetic Clay Liners. Environmental Engineering Research. 2020;**25**(2):163-170. DOI: 10.4491/eer.2019.004

[9] Hadoudi N, Charki A, El Ouarghi H, Mourabit F, Salhi A, Amhamdi H. Sorption of Pentachlorophenol from Aqueous Solution by Acid Activated Bentonite: A comparative study. Egyptian Journal of Chemistry. 2023;**66**(6):353-361. DOI: 10.21608/ejchem.2022.143001.6255

[10] Djebri N, Boutahala M, Chelali N. E, Boukhalfa N, Larbi Z. Adsorption of bisphenol A and 2, 4, 5-trichlorophenol onto organo-acid-activated bentonite from aqueous solutions in single and binary systems. Desalination and Water Treatment 2017;66:383-393. DOI: 10.5004/dwt.2017.20220

[11] Zhou L, Pan S, Chen X, Zhao Y, Zou B, Jin M. Kinetics and thermodynamics studies of pentachlorophenol adsorption on covalently functionalized Fe_3O4@SiO2–MWCNTs core–shell magnetic microspheres. Chemical Engineering Journal. 2014;**257**:10-19. DOI: 10.1016/j.cej.2014.07.060

[12] Hameed BH, Tan IAW, Ahmad AL. Adsorption isotherm, kinetic modeling

and mechanism of 2, 4, 6-trichlorophenol on coconut husk-based activated carbon. Chemical Engineering Journal. 2008;**144**(2):235-244. DOI: 10.1016/j.cej.2008.01.028

[13] Kakavandi B, Jahangiri-rad M, Rafiee M, Esfahani AR, Babaei AA. Development of response surface methodology for optimization of phenol and p-chlorophenol adsorption on magnetic recoverable carbon. Microporous and Mesoporous Materials. 2016;**231**:192-206. DOI: 10.1016/j.micromeso.2016.05.033

[14] Xu L, Wang Z, Ye S, Sui X. Removal of p-chlorophenol from aqueous solutions by carbon nanotube hybrid polymer adsorbents. Chemical Engineering Research and Design. 2017;**123**:76-83. DOI: 10.1016/j.cherd.2017.04.034

[15] Koumanova B, Peeva-Antova P. Adsorption of p-chlorophenol from aqueous solutions on bentonite and perlite. Journal of Hazardous Materials. 2002;**90**(3):229-234. DOI: 10.1016/S0304-3894(01)00365-X

[16] Ji Y, Zhong H, Chen P, Xu X, Wang Y, Wang H, et al. Single and simultaneous adsorption of methyl orange and p-chlorophenol on organo-vermiculites modified by an asymmetric gemini surfactant. Colloids and Surfaces A: Physicochemical and Engineering. 2019;**580**:123740. DOI: 10.1016/j.colsurfa.2019.123740

[17] Shankar A, Kongot M, Saini V. K., Kumar A. Removal of pentachlorophenol pesticide from aqueous solutions using modified chitosan. Arabian Journal of Chemistry 2020;13(1):1821-1830. DOI : 10.1016/j.arabjc.2018.01.016

[18] Zhou Q, Wang Y, Xiao J, Fan H. Adsorption and removal of bisphenol A, α-naphthol and β-naphthol from aqueous solution by Fe3O4@ polyaniline core–shell nanomaterials. Synthetic Metals. 2016;**212**:113-122. DOI: 10.1016/j.synthmet.2015.12.008

[19] Yang Q, Gao M, Luo Z, Yang S. Enhanced removal of bisphenol A from aqueous solution by organo-montmorillonites modified with novel Gemini pyridinium surfactants containing long alkyl chain. Chemical Engineering Journal. 2016;**285**:27-38. DOI: 10.1016/j.cej.2015.09.114

[20] Hadoudi N, Amhamdi H, Ahari MH. Sorption of bisphenol A from aqueous solutions using natural adsorbents: Isotherm, kinetic and effect of temperature. E3S Web of Conferences. EDP Sciences. 2021;**314**:07003. DOI: 10.1051/e3sconf/202131407003

Chapter 5

Geotechnical Hazards Induced by Montmorillonite Presence

Mihaela Stãnciucu and Iuliana Dogaru

Abstract

Often referred to as expansive soils in international literature, soils with large swelling and shrinking potential, are defined as natural materials that exhibit volume variations related to variations of moisture. Most professionals consider that the expansive manifestations are related to the mineralogic composition, especially with the presence of the smectite class of minerals, more specifically with the montmorillonite. Despite sustained worldwide studies over decades, the direct and firm correlation between the number of clayey minerals and expansive properties has not reached a conclusive form, and the behavior of expansive soils remains still unrevealed and exhibits unexpected features under moistening or drying conditions in natural habit, or in relation with infrastructure works. This chapter presents the results of an extended and complex geotechnical investigation of expansive soils which concludes with the validation of a simple procedure of identification of the areas where the swelling behavior of soils may exceed the equilibrium of the geological structure and produce a variety of effects such as lumps or landslides in areas with sloping terrain.

Keywords: expansive, swelling, shrinkage, volume change, smectite

1. Introduction

Expansive soils are implicitly pointed out associated with large damage to onshore or offshore, over, or underground infrastructures projects in more than 60 countries, on all continents except the Antarctic one, often after long periods of drought, heavy rains, freezing, or unfreezing.

The broadening of expansive soils is usually incompletely presented in geologic or pedologic maps, only for the superficial parts of the ground associated with highly over-consolidated claystone and clay shale [1].

Despite worldwide efforts to depict, understand, and predict these peculiarities, annual damages related to geomechanical processes produced by variations of water content in expansive soils are reported all over the world.

For instance, in UK-£400 million, [2], or in the USA- $15 billion, [3]. In the UK, or in China where expansive soil covers more than one hundred thousand square kilometers, expansive soil behavior is stated as one of the most dangerous geological hazards that affect large and diverse engineering projects [4–6].

Even if the specific influence of these soils is not yet quantified in Romania, large areas and important infrastructures are also affected mainly by landslides.

This generally unfavorable situation is the result of the cumulation of various factors related to (i) incomplete knowledge of these soils in the various conditions of occurrence, both as mineralogical nature and sedimentologic structure; (ii) poor understanding of geotechnical behavior and its limits; (iii) insufficient sampling and specific testing in large projects which may put in evidence the presence and the specific properties of these soils, and finally (iv) the inconsistence bond between geotechnical and hydrogeological investigations.

2. Characterization of expansive soils

2.1 The mineralogical nature

The swelling processes are always associated with the presence of clay minerals. Also named hydrous phyllosilicate, the clay minerals are divided into several classes based on the specific structure composed of silica, alumina, and water with variable content of inorganic ions like Mg^{2+}, Na^{+}, and Ca^{2+}, [7].

There are four major groups of clay minerals that are present in soils. Among, them the layer group is the most important one which comprises: type 1:1 (kaolinite, halloysite, serpentine); type 2:1 (smectite, vermiculite, illite, mica), and type 2:1:1 (chlorite).

The largest swelling potentials are related to the presence of minerals from the smectite family, which contains saponites, montmorillonite, and bentonite, and absorb the largest quantities of water between clay sheets [7, 8].

Physico-chemical characteristics of minerals that dictate the amplitude of swelling potential are in order of impact: average specific surface area, particle thickness, interlayer space, and cation exchange capacity. **Table 1** presents these characteristics specific to some usually encountered clay minerals.

Every particular soil layer is a mixture of clay minerals with other particles of inert mineral type (quartz, feldspar, etc.), a fact that triggers a unique swelling and shrinking behavior. It is considered that clays which are geological materials of great mineralogical variety, may exhibit swelling properties if containing mineral particles more than 50% with less than 2-micron size.

Type	Clay mineral	Particle thickness	Basal spacing	Cation exchange capacity at pH 7	Specific area	Swelling potential
		(nm)	(Å)	(milliequivalent /100 g)	(m^2/g)	
2:1	Smectite-Montmorillonite	2	9.8–20	80–120	40–800	High
2:1	Vermiculite	—	10–15	100–150	760	High
2:1	Illite	10	10–13	10–10	80–120	High
1:1	Kaolinite	100	7.2	3–15	5–40	Almost none
2:1:1	Chlorite	—	14	10–40	10–55	None

Table 1.
Physico-chemical characteristics of minerals [7, 9].

It is concluded [9] that usually, the presence of montmorillonite indicates large probabilities of producing reversible processes, swelling or shrinking, under moisture variation, while soils with a predominance of kaolinite or illite exhibit at drying, in the first instance, an initial large volume decrease and subsequently limited swelling on rewetting.

2.2 Micro and macroscale "symptoms" of expansive soils

In the microscale register, swelling and shrinkage mechanisms of expansive soils may be explained by two major theories: crystal swelling of clay mineralogy in which the water molecules get combined with the cations associated with mineral flakes to form hydrated ones and a diffuse double layer of colloidal chemistry in which around negatively charged clay minerals hydrated ions and polar water molecules are firmly adsorbed and form a fixed layer and a thick diffusion layer thus widening the spacing between mineral particles [10]. During the swelling process, the inter-clay bonds are weakened or broken, the layer structure of clay minerals is deformed, and in the macroscopic field, the strength of the soil decreases.

The magnitude of the swelling-shrinking potential is strictly dependent on the types and the percents of clay minerals and is also directly related to the initial saturation degree, the initial void ratio, or the geologic vertical stress [11].

Over decades, the specific geomechanical behavior of expansive soils has been characterized as dramatic either by:

i. the total collapse of soil structure after a single swell/shrink cycle [12, 13];

ii. the decrease of shear strength from 3 to 5.5 times [14];

iii. the decline with 70–90% of shear resistance, mostly by the reduction of effective cohesion, sometimes very close to or equal to zero [15, 16].

Moreover, the shrinking processes should not be considered reversible, since cracks created after shrinkage may not always completely close up after rewetting, often only deepening the watering front together with the sediment deposition [17].

The area of manifestation of these processes is predominately located at the surface of the terrains, wherever is taking place the seasonal variation of underground water [4], on depths less than 5–6 m [6], according to climate particularities, drying/rehydration regime and mostly to variation of temperature.

Usually, swelling of expansive soils is considered only in the vertical direction, but it has been proved that lateral swelling pressure develops additional stress to the lateral earth pressure on retaining walls which may increase at the bottom of the wall equal to 1.3–4 times the overburden [18], and reduce the bearing capacity of piles [1, 19].

Nonetheless, similar processes may be present at larger depths, by the water flow through more permeable layers (in contact with expansive soils) charged from distant front-loading sections [20].

2.3 Geotechnical identification tests

Various procedures for the identification of expansive soils may be established following the international literature, national codes, or professional guides, which are based on a large variety of parameters, indices, or techniques, however, none of those are comprehensive and worldwide accepted. Thus, in countries with predominantly arid climates, the geotechnical methodology for expansive soil characterization is based on shrinkage tests, while in temperate climates swelling tests prevail.

Jones and Jefferson, [4] differentiate the following laboratory testing classes: (i) swelling tests, which may be divided into swelling strain and swelling pressure tests; (ii) index tests based on basic parameters such as liquid limit, plastic limit, plasticity index (usual or modified); (iii) oedometer-based methods, which may be free swell tests or constant volume tests; (iv) suction-based tests which use soil-water characteristic curves; (v) mineralogical tests.

Nelson et al. [1], define three classes of identification methods: (i) based on physical properties – plasticity, free swell test, potential volume change, expansion index, linear extensibility, and standard absorption of moisture content; (ii) based on mineralogical composition–X-ray diffraction, differential thermal analysis, and electron microscopy; (iii) based on chemical analyses – cation exchange capacity, specific surface area, and total potassium content.

Based on the above-defined tests, various schemes of characterization and classification have been developed worldwide, which may be generally categorized into four groups [4]: free swell, heave potential, degree of expansiveness, and shrinkage potential.

Romanian normative [21] propose a scheme of characterization based on a particular collection of parameters: colloidal fraction (A2μ), plasticity index (Ip), activity index (I_A), plasticity criterion (Cp), free swelling (U_L), shrinkage limit (w_s), volumetric shrinkage (Cv), maximum wetting heat (q_{max}), moisture content at 15 bar suction (w_{15}), and swelling pressure (p_u).

3. Geotechnical hazard induced by montmorillonite presence

3.1 Geologic and hydrogeologic regional frame

As we underline above, swelling processes may be active at larger depths, by the water flow through more permeable layers (in contact with expansive soils) charged from distant front-loading sections.

This is the case of a large structure attributed to the Upper Pliocene-Lower Pleistocene named "Cândeşti Layers", in which expansive soil strata are inconsistently wetted at large and variable pressures.

The geological unit we refer to is deposed over the Getic Depression and Moesian Platform units, on the outer side of the Carpathian Arch (Oriental, Curvature, and Meridional) [22].

It has a thickness that may exceed 250 m and an irregular shape with widths starting from 6 to 8 km in the Curvature area, to more than 80 km in the western Meridional Carpathian as it is presented in **Figure 1** [23].

The "Cândeşti Layers" is considered a multilayer-aquifer (with terms from Jurassic to Upper Pleistocene) with a feeding front in the North end at the contact with the Carpathian Chain.

The main underground flowing directions are from Nord to South, and the hydrogeological characteristics are hydraulic conductivities in the order of about 100 m/day and transmissivity of less than 1000 m^2/day.

The hydrodynamic regime is defined by high pressures (up to 40 Bars), generated by deeper regional aquifers, with which it is in connection. Often in places, the

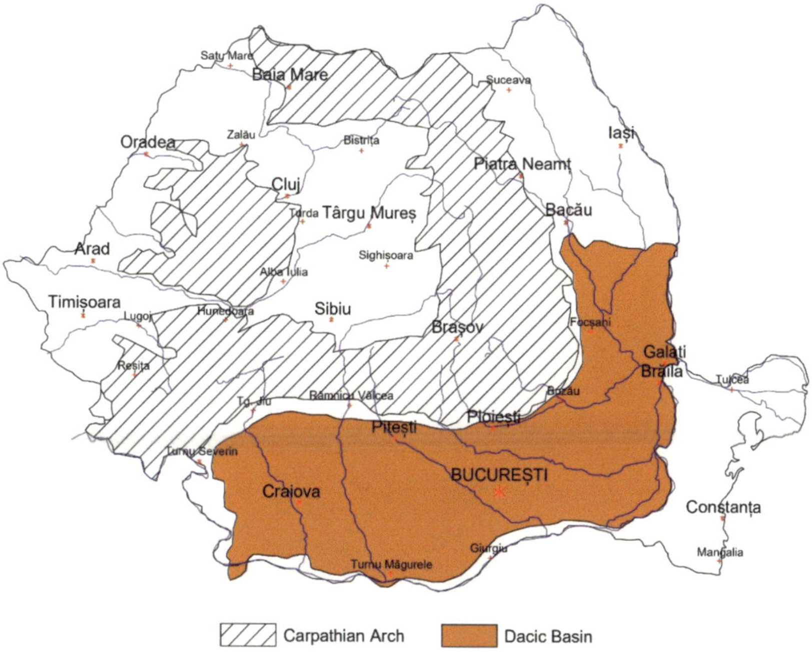

Figure 1.
Extent of the Artesian Dacic Basin [23].

package of multilayer-aquifer, defined also as an "Artesian Dacic Basin", may function under a pressure regime that surpasses the surface of the terrain, causing the watering from the bottom up of the clayey aquitard layers [23].

In this geological and hydrogeological context, the swelling hazard emerges from the rhythmic sedimentation regime which is materialized in alternant sequential layers separated in two main terms: the course one (gravels and sands) and the fine one (expansive clays and fine sand), disposed in the peculiar geologic structure which is presented in **Figure 2**.

3.2 Geotechnical and mineralogical characterization of "Cândeşti layers"

3.2.1 Geotechnical identification

The most common identification tests performed worldwide are grain size distribution and plasticity (Atterberg) limits, which have been executed on the entire set of samples with expansive behavior.

The results of identification tests, presented in **Table 2**, have been divided into three classes according to the plasticity index: class C1- of medium plasticity; class C2- of high plasticity, and C3 of very high plasticity.

3.2.2 Expansive properties

Among the many geotechnical properties supposed to characterize the expansivity of soils, we chose to analyze the most representative ones: free swelling (U_L) and swelling pressure (p_u) in relation to colloidal fraction (A2μ) and plasticity index (Ip).

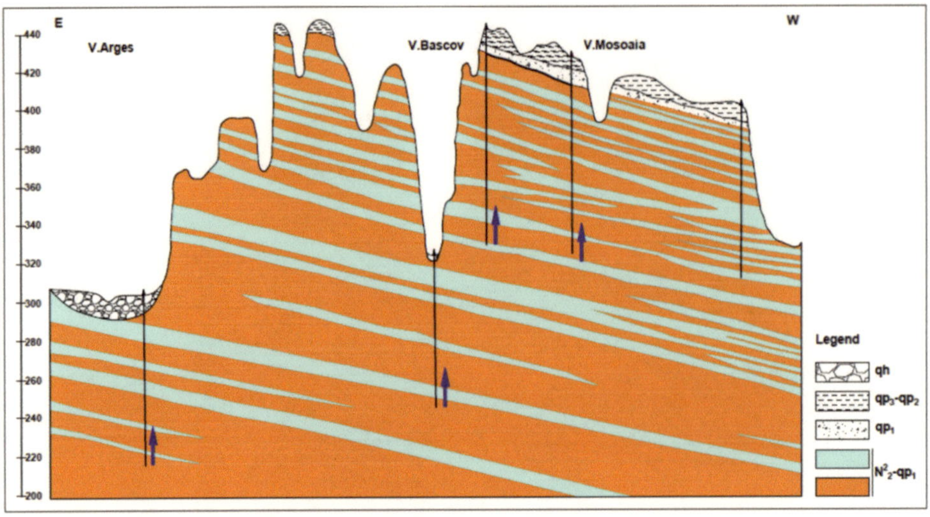

Figure 2.
Exemplification of the mechanism of ascensional watering of expansive soils. Specific geologic cross section of Artesian Dacic Basin [24].

Class of plasti-city	C1 (Ip < 20%)		C2 (20% ≤ Ip < 35%)		C3 (Ip ≥ 35%)		Total no of samples	[%]
	No of samples	[%]	No of samples	[%]	No of samples	[%]		
Cl	—	—	5	2	311	75	316	48
si Cl	1	4	91	41	65	16	157	24
sasi Cl	21	75	73	33	4	1	98	15
sa Cl	3	11	50	22	32	8	85	13
cl Si	—	—	4	2	—	—	4	<1
cl Sa	3	11	—	—	—	—	3	<1
Total	28	100	223	100	412	100	663	100

Table 2.
Results of geotechnical identification of soils.

3.2.2.1 Free swelling (U_L)

It is defined as the increase in the volume of a soil sample, without any external constraints, when submerged in water, the range of values of this parameter may point to soils less active ($U_L < 70\%$), with medium activity ($70\% < U_L < 100\%$), active ($100\% < U_L < 140\%$), or very active ($U_L > 140\%$), in relation with water, according to Romanian normative [21].

In **Figure 3** the values of U_L are plotted in different colors, according to the class of plasticity to which they are attributed. It presents the variation in depth of U_L and reveals that according to this criterion, 35% of samples have medium activity, 28% are active and 24% are very active. Concerning the last category of very active samples, we mention that most part of those (79%) have $U_L < 300\%$, and the highest recorded value is $U_L = 603\%$.

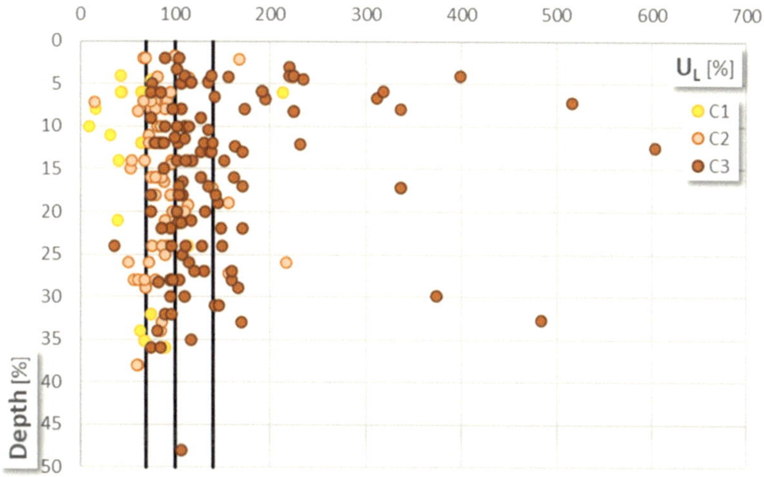

Figure 3.
Variation of U_L in-depth.

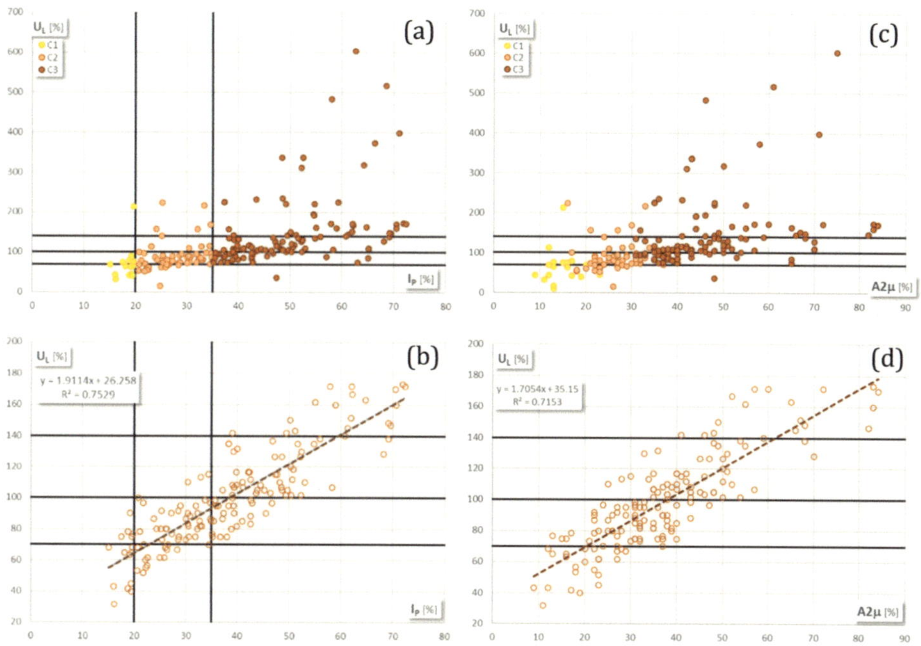

Figure 4.
(a and b) Variability between U_L and Ip. (c and d) Variability between U_L and A2μ.

Figure 4a and **c** reveals poor correlations between U_L and Ip or A2μ unfulfilled by unexpected great values of U_L registered in all plasticity classes. After excluding the highest values of free swelling, ($U_L > 190\%$), from both ranges, the variabilities UL = U_L(Ip) and U_L = U_L(A2μ) have settled in good linear correspondences.

3.2.2.2 Swelling pressure

It is defined as the external pressure required to consolidate a sample to its initial void ratio, after the expansion due to water submersion, or the pressure required to hold the soil at constant volume when water is added, the swelling pressure may be tested by two [1] or three methods [25].

The most common are the conventional consolidation test which usually provides upper bound values and the constant volume method. Less used is the method of equilibrium of void ratios which gives lower values. In this research, the conventional consolidation test has been used.

Class of swelling pressures		C1		C2		C3		Total no of samples	[%]
		No	[%]	No	[%]	No	[%]		
(I)	$p_u < 200$ kPa	28	100	203	91	254	62	485	73
(II)	$200 \le p_u < 400$ kPa	—	—	16	7	114	28	130	20
(III)	$p_u \ge 400$ kPa	—	—	4	2	44	11	48	7

Table 3.
Classification system according to swelling pressure results.

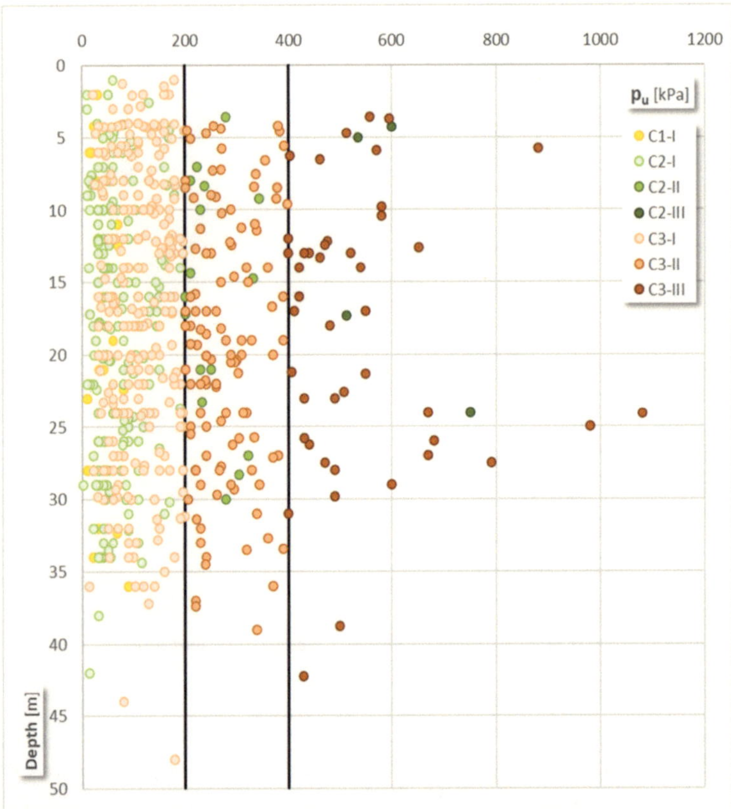

Figure 5.
Variation of p_u in depth.

For smooth processing of the data, an additional classification of samples has been made according to the range of values of this important parameter. The new system overlaps the precedent one and is presented in **Table 3**.

As one can visually observe from **Figure 5**, about three-quarters of the total amount of samples (73%) present swelling pressures under 200 kPa (class I). Samples of medium and high plasticity (C1-C2) are entirely enclosed in this pressure class (100–91%), while samples of very high plasticity (C3) present this range of swelling pressures at 62%.

The highest values of swelling pressure grouped in class (III) represent 7% of the samples, which belongs mainly to class (C3). Concerning the class (III) of swelling pressure, we underline that most part of these samples (80%) have $p_u < 600$ kPa, and the highest value recorded exceeds 1000 kPa.

Figure 6a and **c** presents the correspondence between p_u and Ip, respectively p_u and A2μ, depicted according to both systems of classification. Once again unpredictable, high values of swelling pressure, encountered in classes C2-II, C2-III, and C3-III make the correlations weak, and it still remains poor even after excluding some unusual values ($p_u > 400$ kPa).

Finally, for some of the samples that have been tested both for free swelling and for swelling pressure (**Figure 7a**), a medium-strong correlation has been established after excluding values with $p_u > 400$ kPa (**Figure 7b**).

3.2.3 Mineralogical identification

As part of the specific characterization of Cândeşti Layers, several mineralogical analyses have been executed which specified that the swelling minerals prevailed in all tested samples as is presented in **Table 4**.

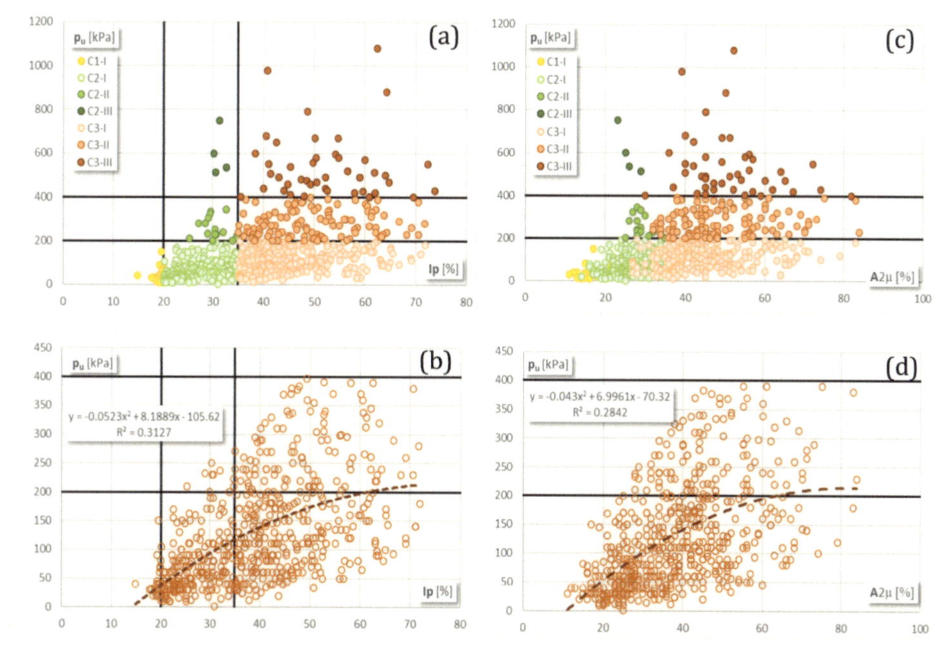

Figure 6.
(a and b) Correspondence between p_u and Ip. (c and d) Correspondence between p_u and A2μ.

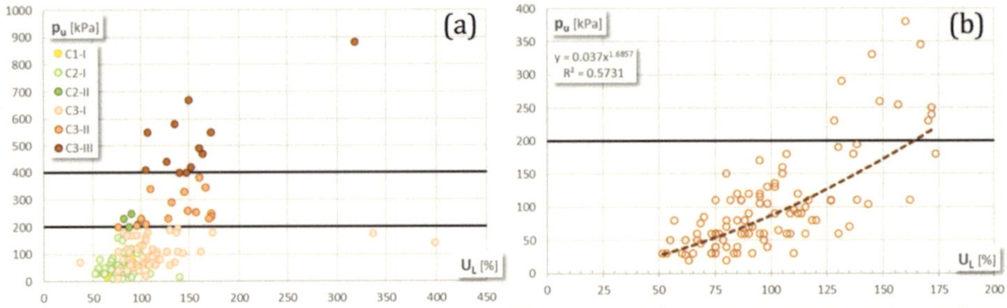

Figure 7.
Correspondence between p_u and U_L.

Class	Swelling pressure classes pu (kPa)	Samples (10)	Percents of minerals (%) X-ray diffraction			
		Number	Smectite	Illit	Kaolinit	Chlorit
II	$200 < p_u < 400$	4	70–77	18–25	4–6	—
III	$p_u > 400$	6	63–79	17–32	2–4	3–8

Table 4.
Mineralogic characterization of Cândeşti Layers samples.

The result of these mineralogical analyses confirms the presumption of the presence of hydrous phyllosilicate in expansive soils (pressure classes II and III). More than that, the fact that smectite class of clay minerals – which includes the montmorillonite - prevails in all tested samples explains the unusual values of geotechnical parameters U_L and p_u recorded in this geologic unit.

3.3 Quantification of geotechnical hazard of "Cândeşti Layers"

Starting from the hydrodynamic characterization of the multi-layer aquifer "Cândeşti Layers" exposed above (in §3.1.), the assessment of the geotechnical risk of swelling of aquitards expansive terms submitted to pressures of underground waters raised at the regional scale, up to 40 Bars, conducted to postulate that risk of moistening by vertical ascendant drainage is high and affects the whole area of expansion of this unit.

By considering the wetting zone in acceptance of [26], as the zone in which water contents may increase by supply from external sources, including by capillarity rise, the peculiar sedimentologic and hydrogeologic architecture described above, conduct in geotechnical terms to the extension in depth of it, up to the thickness of Cândeşti Aquifer.

In consequence, we began to evaluate the depth of the potential heave as the maximum depth at which the swelling pressure of the soil equals or exceeds overburden vertical stress [26], a quantity that we will note Δ_{sw}, according to (Eq. 1).

$$\Delta_{sw} = \sigma_g - p_u \quad (1)$$

Figure 8a presents the repartition in depth of the values of (Δ_{sw}), which allows us to observe that for a large part of tested samples, this difference is positive $(\Delta_{sw} > 0)$,

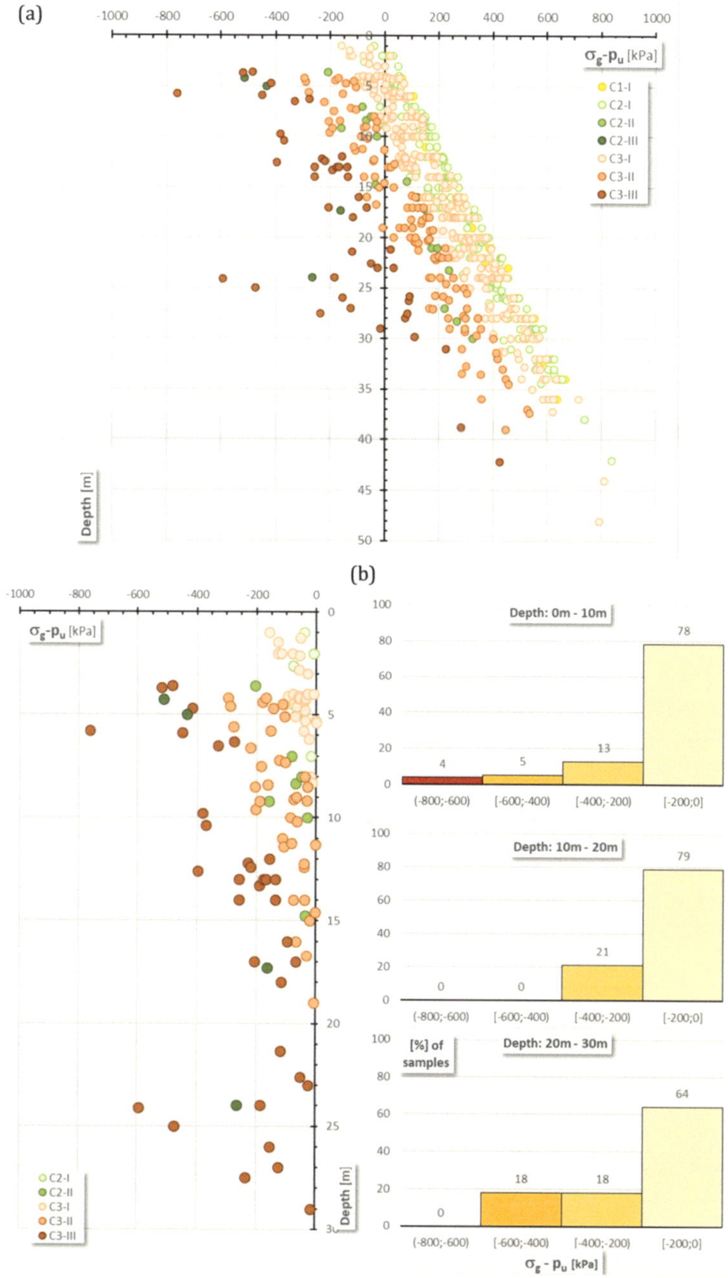

Figure 8.
(a) Variation in depth of the potential heave of Cândeşti Layers. (b) Detail of negative values of the potential heave.

and thus, the presence of expansive soils in foundation terrains of shallow or deep foundations is outside of this geotechnical hazard. Nevertheless, 9% of C2 and 25% of C3 (summing up 19% of all expansive samples, regardless of swelling pressure class), reveal negative values of $(\Delta_{sw} < 0)$, which are presented in detail in **Figure 8b**.

One can observe that the main part of these negative differences (77%) are less than -200 kPa while 9% of samples present values between -800 kPa and -400 kPa and are located in the first 10 m depth from the surface of the terrain. The whole range of negative values is extended in depth up to almost 29 m, which is in consequence the thickness of the potential heave.

In cases of special interest and limited extension, a detailed design must be based furthermore on more specific laboratory trials of every expansive layer situated in the active zone, such as consolidation-swell tests and constant volume tests, which will allow a proper settlement calculation [1, 20].

In cases of broad infrastructure projects that run through geological structures exposed to swelling hazards on large depths, such as the one presented above, the calculations of (Δ_{sw}) values must be considered as a qualitative evaluation and a fundamental first step in the process of assessment of this geotechnical risk. The assessment process once triggered in specified areas, must be continued with denser and specific tests in order to reveal the precise extension of these types of soils.

This simple procedure based on the most common geotechnical tests, executed in the preliminary stages of investigations, may point not only to the extension of expansive soils but also to the areas where the swelling behavior of soils may exceed the equilibrium of the geological structure and produce a variety of effects such as lumps or even landslides in areas with sloping terrain.

As an exemplification of this procedure, **Figure 9a** presents a vertical section alongside an infrastructure route that crosses a zone located in the Artesian Dacic Basin, with the distribution of isobars of swelling pressures recorded at different depths. Details represented in **Figure 9b**, allow the visualization of the specific extension in depth of zones of potential heave, and consequently the extension of the swelling geotechnical risk.

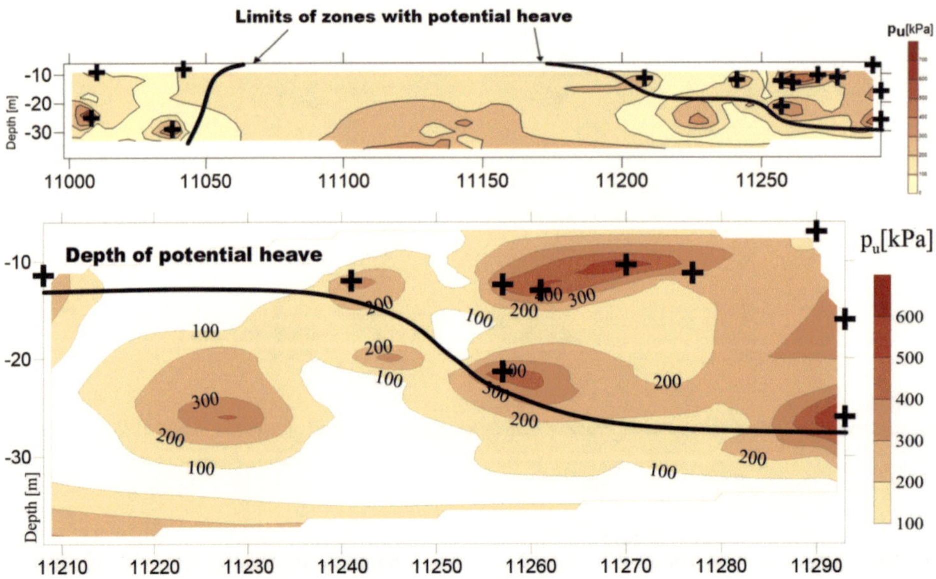

Figure 9.
(a) Chart of the distribution of zones of potential heave alongside an infrastructure route. (b) Detail with extension in depth of zone of potential heave.

4. Conclusions and outlook

In the worldwide context of poor mineralogical testing associated with geotechnical investigations required for large infrastructure objectives, the random presence of montmorillonite inside sedimentary layers represents a strong geotechnical hazard, that may produce large damages of shallow or deep foundations.

The paper presents such a study case in which exceptional swelling pressures have been revealed by numerous specific geotechnical investigations (663 expansive soil samples) to whom they have added only several mineralogical analyses by X-ray diffraction.

In order to characterize the swelling properties of soils, two specific parameters have been analyzed (U_L and p_u) in correlations with two general parameters (Ip and A2μ), in the frames of two classification systems. The main results may be synthesized as follows:

- concerning the free swelling parameter:
    - the highest recorded value is $U_L = 603\%$;
    - the most part of samples (76%) presented values $U_L < 140\%$;
    - variabilities $U_L = U_L(Ip)$ and $U_L = U_L(A2\mu)$ reveal weak correlations due to unexpectedly great values of U_L registered in all plasticity classes, which have settled in good linear correspondences ($R^2 = 0.71–0.75$) after excluding from both ranges the highest values of free swelling ($U_L > 190\%$);
- concerning the swelling pressure parameter:
    - the highest value recorded is $p_u = 1080$ kPa, followed by 7% of samples with $p_u > 400$ kPa;
    - variabilities $p_u = p_u\ (Ip)$ and $p_u = p_u\ (A2\mu)$ reveal weak correlations due to unexpected great values of p_u, which remains in poor correspondences ($R^2 = 0.28–0.31$) even after excluding from both ranges the highest values of swelling pressures ($p_u > 400$ kPa);
- for some of the samples that have been tested both for free swelling and for swelling pressure (133 samples), a medium-strong correlation $p_u = p_u\ (U_L)$ with $R^2 = 0.57$, has been established after excluding values with $p_u > 400$ kPa;
- the lack of strong and "natural" correspondence between swelling parameters (U_L, p_u) and general parameters (Ip, A2μ) is due to the randomness presence of hydrous phyllosilicate minerals, namely smectite class which includes the montmorillonite, a well-known responsible for the expansion of clayey soils;
- for nine samples collected from swelling pressure classes II and III the percentual amount of smectite class was in great amount, between 63% and 79%;
- given the fact that the distribution of montmorillonite across the entire sedimentary structure of various thicknesses is randomly, and in the context of peculiar hydrogeologic architecture of Cândeşti Aquifer (multistate-aquifer with high pressiometric regime), the probability of wetting of expansive soils at various

depth is high; all these findings drive to a general characterization of these soils in natural conditions or in relation with great infrastructure projects, as having a hazardous geotechnical behavior;

- finally, the evaluation of the depth of the potential heave according to [26], offers quantitative spatial indications regarding the active zone in which the terrains may be affected by swelling/shrinking processes.

Based on the previous conclusions we suggest several directions to conduct proper investigations in the area of occurrence of this particular swelling formation, in order to reduce the degree of uncertainty of knowledge, and thus to reduce the geotechnical hazard related to the presence of montmorillonite in geological structures by:

- intensive boosting the research activities of swelling potential mapping at the level of every country involved in large infrastructure projects, regardless of the source of funding;

- the swelling potential mapping activities should:

  - be executed by interdisciplinary teams composed of geologists specialized in clay mineralogy, engineering geologists, hydrogeologists, and civil engineers;

  - start from an up-to-date geologic map, combined with a regional hydrogeologic map containing the regional underground water corps at convenient scales (1: 50000 to 1:15000) in order to be of practical use, not just scientific;

  - be based on geological and geotechnical investigation reports executed in the concern area, which must be accessed based on the principle of public interest;

  - be focused not only on the extent of expansive soils in 3D directions but also on the limits of variation of superficial underground water table levels and of piezometric regime of deeper aquifers in hydraulic contact with expansive layers;

  - be accompanied either by a digital 3D model with the extent of expansive soil layers combined with upper and lower limits of water table/piezometric levels, or by a dense grid of cross and longitudinal sections extended in depth up to the level of the deeper potential heave reported in the area;

  - contain also delimitations of the areas of landslides (regardless of the stage of evolution) and superficial damages to buildings (civil, industrial, or infrastructures);

- community standardization of types, numbers, and depth of geological and geotechnical investigations regardless of the purpose, in areas of expansive clay occurrences;

- community standardization of types, numbers, and density of laboratory geotechnical and mineralogical analyses that must lead to a correct and complete characterization of expansive layers, according to the stage of further investigations (preliminary, design, detail, or monitoring).

Swelling potential maps should be exposed and promoted in an open access regime, but in the meantime must be enclosed and endorsed by national standard normative and become mandatory to consider in order to prevent further damage to individual or large infrastructure investments.

Acknowledgements

The authors are grateful to the company Geotesting CI, Bucharest, Romania, for the permission to use the laboratory data included in this paper.

Author details

Mihaela Stănciucu* and Iuliana Dogaru
University of Bucharest, Romania

*Address all correspondence to: stanciucumihaela@yahoo.com

References

[1] Nelson JD, Chao KC, Overton DD, Nelson EJ. Foundation Engineering for Expansive Soils. USA: John Wiley & Son; 2015. 385 p

[2] Driscoll R, Crilly M. Subsidence Damage to Domestic Buildings. Lessons Learned and Questions Asked. London: IHS BRE Press; 2000. p. 32

[3] Nelson JD, Miller DJ. Expansive Soils: Problems and Practice in Foundation and Pavement Engineering. USA: John Wiley & Son; 1992. 259 p

[4] Jones LD, Jefferson I. Expansive soils. In: Burland J, editor. ICE Manual of Geotechnical Engineering. Vol. 1, Geotechnical Engineering Principles, Problematic Soils and Site Investigation. London, UK: ICE Publishing; 2012. pp. 413-441

[5] Gaspar TAV, Jacobsz SW, Heymann G, Toll DG, Gens A, Osman AS. The mechanical properties of a high plasticity expansive clay. Engineering Geology. 2022;**303**. 11 p., 106647. Availlabel from: www.elsevier.com

[6] Xiao J, Yang H, Zhang J, Tang X. Surficial failure of expansive soil cutting slope and its flexible support treatment technology. Advances in Civil Engineering. 2018;**1609608**. 13 pp

[7] Kumari, N. and Mohan, C. (2021). Basics of Clay Minerals and Their Characteristic Properties. Chapter in Clays and Clays Minerals. Ed. G. M. Do Nascimento. IntechOpen, London, UK; 2021. 30pp. 10.5772/intechopen.97672

[8] Mitchell JK, Soga K. Fundamentals of Soil Behavior. 3th ed. New York: Wiley; 2005. 560p

[9] Yong RN, Masashi Nakano M, Pusch R. Environmental Soil Properties and Behaviour. Taylor Francis Group: CRC Press; 2012. 455 p

[10] Zhou S, Zhou D, Zhang Y, Wang W. Study on physical-mechanical properties and microstructure of expansive soil stabilized with fly ash and lime. Advances in Civil Engineering. 2019. 15 p

[11] Bell FG, Culshaw MG. Problem soils: A review from a British perspective. Proceeding of the Symposium on Problematic Soils. Nottingham. 2001. pp. 1-37. ISBN: 0 7277 3043 6

[12] Bishop AW, Kumapley NK, El-Ruwayih AE. The influence of pore-water tension on the strength of clay. Philosophical Transactions of the Royal Society London. 1975, 1975;**275**(1286):511-554

[13] Alonso EE, Gens A, Josa A. A constitutive model for partially saturated soils. Geotechnique. 1990;**40**(3):405-430

[14] Sherif M, Mazen O, Gergis NS. Behavior of expansive soil during shear. In: Proceedings of the First National Conference on the Science and Technology of Buildings, Khartoum, Sudan. 1984. pp. 557-562

[15] Al-Mhaidib A, Al-Shamrani MA. Influence of swell on shear strength of expansive soils. Proceedings of the geo-Shanghai conference advances in unsaturated soils. Seepage and Environmental Geotechnics. 2006;**148**:60-165

[16] Lan T, Zhang R, Yang B, Meng X. Influence of swelling on shear strength of expansive soil and slope stability. Frontiers in Earth Science.

2022;**10**:849046. DOI: 10.3389/feart.2022.849046

[17] Khademi F, Budiman J. Expansive soil: Causes and treatments i-Manager's. Journal on Civil Engineering. 2016;**6**(3):1

[18] Abdollahi M, Vahedifard F. Prediction of lateral swelling pressure in expansive soils. In: Proceedings, Geo-Congress: GeoSystems, Sustainability, Geoenvironmental Engineering, and Unsaturated Soil Mechanics. Reston, VA: ASCE; 2020. pp. 367-376

[19] da Silva Burke TS, Jacobsz SW, Elshafie MZEB, Osman AS. Measurement of pile uplift forces due to soil heave in expansive clays. Canadian Geotechncial Journal. 2022;**59**:2119-2134

[20] Nelson JD, Overton DD, Chao KCG. Design of foundations for light structures on expansive soils. In: California Geotechnical Engineers Association Annual Conference. Pub. Carmel; 2003

[21] NP 126. Normative Regarding Building Foundations on Soils with Large Swelling and Shrinkage. (Roumanian Language). Bucharest: Romanian Ministry of Regional Development and Tourism; 2010

[22] Jipa DC. Paleogeographic evolution of the Dacian Basin: Emergence, development and closure of the basin. In: Jipa DC, editor. Bazinul Dacic. Sedimentary Architecture, Evolution, Control Factors. Ed. Geoecomar, Bucharest. 2006. pp. 17-32

[23] Palcu M, Melinte MC, Jurkiewicz A, Witek G, şi Rotaru, A. Preliminary inventory of aquifer structures in the southern part of Romania. Rew. Geo-Eco-Marina. 2008;**17**:7-16

[24] Feru M, Scafă C, Szabo N, Liteanu E, Pricăjan A, Andreescu I, et al. Hydrogeological map, sc.1:100 000, Sheet 34d. Piteşti, Romanian: Institute of Geology and Geophysics. Bucharest; 1980

[25] Sridharan A, Rao AS, Sivapullaiah PV. Swelling pressure clays. Geotechnical Testing Journal. GTJODJ. March 1986;**9**(1):24-33

[26] Nelson JD, Overton DD, Durkee DB. Depth of wetting and the active zone. Expansive clay soils and vegetative influence on shallow foundations. ASCE Geotechnical Special Publications, 2002. 2001;**115**:95-109

Chapter 6

Research for Industrial Application of Bentonite-Polymer Material in Ferrous Metallurgy

Daniil Vetyugov and Tamara Matveeva

Abstract

Ferrous metallurgy, in particular the process of pelletizing iron ore concentrates, is one of the main consumers of bentonite clays. The problem with the bentonite binder is well known, and is the increase in silica content (the main harmful impurity) in the roasted pellets as a result of its use. This predetermines long-term interest in the development of new binders that have lower consumption, or do not contain silicon dioxide at all. Increasing the quality characteristics of the binder makes it possible to reduce its consumption, thereby optimizing the chemical composition of the roasted pellets. The results of experimental studies on the palletization of magnetite concentrate from several iron ore plants, different enrichment depths (Fe content = 65–71%), and basicity $(CaO)/(SiO_2)$ = 0.3–1.0 are presented. It has been shown that using the effect of mixing bentonite and a polymer additive on the binding properties of their compound makes it possible to increase the strength characteristics of pellets relative to those in current production (without polymer) and more significantly than when excluding bentonite in the case of its complete replacement with an organic binder. Much attention is paid to studying the influence of the use BPC (Bentonite Polymer Composition) on the metallurgical properties of finished pellets.

Keywords: iron ore concentrate, pelletizing, roasting, bentonite binder, polymer additives, bentonite-polymer composite, iron ore pellets

1. Introduction

The purpose of agglomeration of iron ore concentrates (pelletizing, agglomeration, and briquetting) is to create a product of the same size and shape, intensifying the recovery process. Most iron ore raw materials are processed into cast iron using a blast furnace, which requires raw materials with high reduceability and strength to increase the economic efficiency of operation. Achieving these conditions is ensured by loading the pellets inside the blast furnace without compromising the permeability of the material layer with an upward flow of reducing gas (corresponding porosity) and appropriate porosity of the pellets for the complete access of the reducing gas to the entire volume of the pellet.

A minority of pellets are created for the direct reduction process, which requires easily reducible feedstock with minimal silica content. The strength of such pellets is also important for their transportation and handling and is included in the list of requirements for the production of strong pellets. The strength of roasted pellets affects the amount of material lost due to destruction. In addition, the material is lost in the form of dust due to the abrasion of the pellets (**Figure 1**).

The use of bentonite provides guaranteed strength characteristics of pellets that meet the requirements of transportation, blast furnace process, or metallization [1]. The significant advantage of bentonite is confirmed by the extensive accumulated experience of its use, as well as an extensive array of literary sources and reference data. At pelletizing plants in the Russian Federation, the dosage of bentonite is 0.5–0.8% by weight [2]. Bentonite is a traditional binder for the production of iron ore pellets and consists of fine clays, represented by at least 70% minerals of the smectite group (mainly montmorillonite), which belongs to the subclass of layered silicates and consists, %: 48–56 SiO_2; 11–22 Al_2O_3; 0–5 or more Fe_2O_3; 4–9 MgO; 0.8–3.5 or more CaO; 12–24 H_2O [3]. Among the properties of montmorillonite that determine the industrial value of bentonite: submicron crystal size, layered structure, large surface area (up to 800 M^2/g), significant negative charge of the layer (≈120 mEq/100 g), and associated balancing exchangeable cations located on the surface [4]. These structural features of montmorillonite determine its specific properties, such as binding and sorption capacity and heat resistance.

In terms of volume of consumption, the most important areas of use of bentonite clay in the world are metallurgy, foundry, and drilling. Also, bentonite clays and

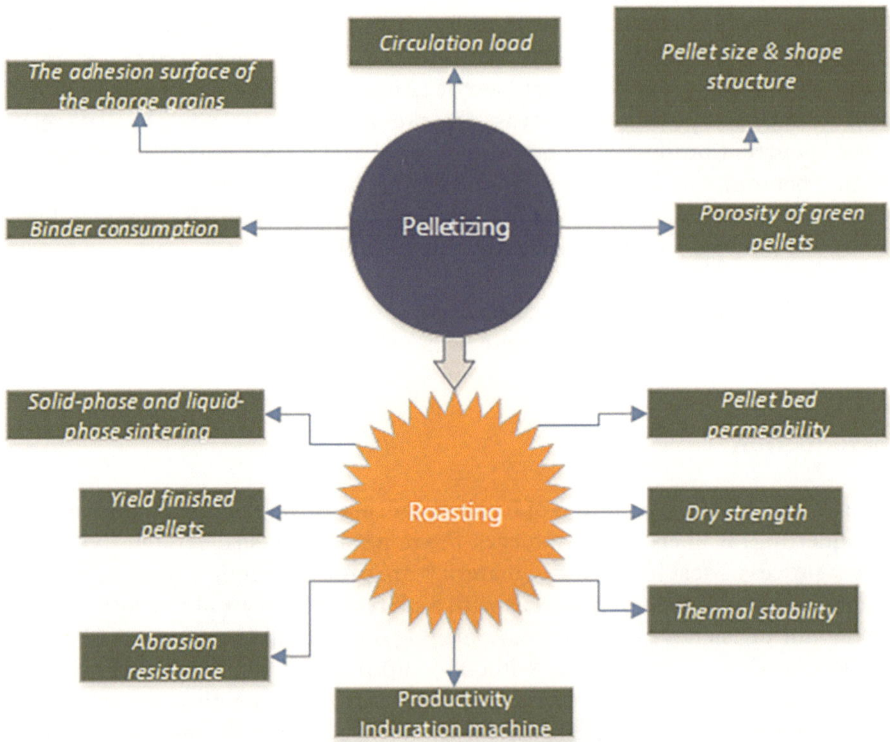

Figure 1.
Parameters depend on the binder used in the production of iron ore pellets.

materials based on them are used in other industries, including waterproofing materials, agriculture, rubber, polymer, paper industries, pharmacology, etc. [5, 6].

The main disadvantage of bentonite is the addition of silica to the pellets [7, 8]. In addition to optimizing the chemical contents (increasing the Fe_{total}% content in the finished pellets), reducing the silica content leads to a decrease in the amount of flux that must be added to achieve certain levels of pellet basicity, which ultimately leads to a decrease in slag volumes in blast furnaces and other units in the production of cast iron. The situation is aggravated, including for calcium bentonites, which are more accessible but less effective and require a higher dosage to achieve equivalent pellet strength [9]. Bentonite clays are divided according to the nature of the exchangeable cations in the interlayer complex of montmorillonite into alkaline (sodium) and alkaline earth (calcium and magnesium) types. Alkaline bentonites have higher technological properties compared to alkaline earth ones due to the fact that alkali metal ions, primarily sodium, have a higher hydration potential [10]. Considering that the swelling of Na-montmorillonites is significantly higher than that of Ca-montmorillonites, to improve the quality indicators, Ca-montmorillonites are converted into sodium form by activation with soda ash [11].

Despite the extensive volume of literature [12–14], the analytical study of the question of how the mechanism of operation of the pelletizing binder is implemented seems insufficient. This is especially true for composite binders consisting of two or more different materials. Research into minimizing binder consumption is clearly necessary to insure a better product and lower operating costs. A rational approach to the choice of binder to obtain strong pellets is also necessary. Even in the worst case, in which the test binder does not have any effect on the heat-strengthening process, having a thorough understanding of the process will allow you to avoid research on initially unpromising binders [15].

A material that is not dispersed throughout the pellet will not be effective as a binder. There are many examples of materials that have been tested as binders but have been ineffective in improving dry pellet strength, which is especially true for non-swelling clays (e.g., attapulgite and kaolinite). For the same reason, insoluble organic substances (e.g., oils) are, in most cases, unpromising as a binder. A material that cannot form bonds with the concentrate particles is also clearly unsuitable for use as a binder. Most industrially used binder bonds through hydroxyl of carboxyl groups, sometimes based on an inorganic base such as Si—OH in bentonites or metasilicates.

It is generally accepted that the successful replacement of bentonite with another inorganic, organic, or composite material will ultimately solve the binder problem [16]. At the same time, it is necessary to distinguish between the binding and strengthening properties of the material used. The uniqueness of bentonite for the production of pellets is largely due to the combination of its binding properties at the stage of raw palletization and its strengthening properties at the sintering stage. Studies conducted with various mono-materials have shown that, most often, materials with excellent binding ability (organic) do not participate in thermal strengthening in any way. And vice versa—inorganic materials (e.g., colemanite, which, due to its property of reducing the sintering temperature, has found popularity among researchers [17]) practically do not contribute to the clumping of raw pellets but significantly intensify the sintering process. This reasoning in itself suggests that the most rational approach to solving the binder problem can be the study of combinations of binders and strengthening materials and, in the best case, achieving a synergistic effect from the interaction of the inorganic and organic parts of the composite material.

The typical mechanism of action of an organic binder is similar to that of bentonite clay, which is water-dispersible and forms bridges between concentrated particles. Inorganic binders exhibit a variety of mechanisms of interaction with concentrated particles and with each other. In addition to regulating the strength of the pellets (crushing strength), the choice of binder affects their dust formation (abrasion strength).

During the roasting of pellets, pyrolytic volatilization of organic binders occurs, while inorganic binders remain dispersed in the pellet volume and affect the sintering stage. Sintering is the main method of strengthening pellets for the blast furnace process [18]. The energy costs for roasting are the highest at a mining and processing plant, and in this respect, they exceed the grinding process. The approximate energy intensity of roasting magnetite pellets is 320–475 MJ/t [19], subject to the use of the best available technologies in the field of process energy efficiency, heat recovery [20], and intensification of thermal strengthening using various additives.

The traditional scheme for producing iron ore pellets by pelletizing the concentrate and further roasting the pellets has fundamental disadvantages. The desire to increase the unit power of units with the principle of palletization of charge materials underlying the production of cast iron has led to the bulkiness and energy inefficiency of the multi-link technological scheme. The idea of pelletizing a charge conflicts with the principle of global energy saving (entropy) since the huge reaction surface of ore materials crushed at the enrichment stage is subsequently reduced by at least a thousand times (**Figure 2**), at the cost of large material and energy costs of sintering and coke production. Heating and melting of agglomerated materials requires a long residence time and, consequently, a large volume of aggregates. In this case, the processes occur close to the state of thermodynamic equilibrium [21].

Large resources are invested in the search and development of a method for the cost-effective production of pellets strengthened without the use of high-temperature firing because, in theory, pellets produced using non-roasting technology will have a lower production cost and a higher final price, given their suitability for carbon-neutral steelmaking technologies. Of the binder options known today, the following are used: Portland cement, glue, sodium metasilicate, and organic compounds. Among the chemical-catalytic methods of hardening are known: carbonization of lime at 100–105°C, hardening of magnesium chloride, and hydrothermal treatment under pressure. Since bentonite is the carrier of the strength of dry pellets in current production,

Figure 2.
The path of iron from the bowels of the earth to the blast furnace.

it can be assumed that in non-roasting technology, bentonite will be represented by a larger or smaller part of the strengthening mixture.

The paper presents the results of experimental studies of the pelletizing process using bentopolymer compositions (BPCs). Since the strength of fired pellets necessary and sufficient for the blast furnace process is successfully achieved in most industries, the key task of the research was, in addition to finding ways to reduce bentonite consumption, to improve the metallurgical properties of the pellets. It is taken into account that the strength of roasted pellets characterizes the completeness of the oxidation process—solid-phase sintering. In cases where the strength of pellets from the use of BPC at the same consumption decreased significantly, no further studies of such a binder were carried out. The nature of the relationship between metallurgical properties (provided mainly by total porosity, pore morphology, and, accordingly, reducibility) and the strength of pellets is still of interest, since practically obtained data often differ from theoretical assumptions. This is largely due to the fact that the size of recrystallized hematite grains and the temperature limits for the formation of the ore binder at the stage of solid-phase hardening are not the same for iron ore concentrates from different deposits.

2. Research methods and objects

This paper presents the summary results of several studies conducted in 2021–2023 on charge material from various iron ore enterprises in the Russian Federation and Kazakhstan.

The main component of bentopolymer composition was alkali-activated bentonite clay from the 10th Khutor deposit (Republic of Khakassia) [22], a key supplier of bentonite for the production of iron ore pellets in the Russian Federation. For the use of bentonite clay in the production of pellets at iron ore mining and processing plants, there are a number of regulated indicators that characterize the quality of the clay. For example, regulated indicators may include, in addition to the swelling index and effective viscosity, the content of montmorillonite (**Table 1**). The rheological properties and swelling of bentonite raw materials, in addition to the mineral composition, are influenced by the crystal-chemical (octahedral lamellar composition, type of counterion, and layer charge) and morphological (size and shape) parameters of montmorillonite.

The content of montmorillonite was determined using the method of determining the cation exchange capacity by adsorption of the methylene blue dye (MB), described both in Russian standards [23] and in foreign literature [24].

The content of montmorillonite directly correlates with the degree of swelling and effective viscosity. Analysis of clay mineral content is a non-trivial task and requires specialized equipment and the work of experienced analysts. The micron and

The name of indicators	Measure	Value, no less	"10th Khutor" 5th layer center
Effective viscosity	mPa s	28	41.44
Swelling index	ml/2 g	30	33
Content montmorillonite	%	53	62.4

Table 1.
Regulated quality indicators of bentonite clay for iron ore pelletizing.

submicron sizes of aggregates of clay particles, the complex structure of crystallites, the presence of mixed-layer formations, and numerous impurities make clay diagnostics difficult. The main methods for determining the content of montmorillonite are related to the calculation of the adsorption value of organic dyes [25].

For use as the polymer part of the composite binder, a variety of substances were tested and selected from the following considerations: priority was given to organic compounds that demonstrate the effect of increasing rheological properties and regulating filtration when interacting with bentonite. We are talking about polymers that are successfully used to improve bentonite drilling fluids. Among them are polyacrylamides, carboxymethylcellulose, and xanthan gum.

Possible schemes for producing composite binders involve dry mixing of the polymer and bentonite parts in one or a combination of several different ways, ranging from joint extrusion under pressure and joint grinding, to simple mixing. Considering the opinion that mechanical activation of charge components during the production of pellets in terms of resource-saving is unacceptable [26], pilot batches of composite binders were obtained by mixing. Mixing is an important production stage in preparing the charge for palletization. Many works are devoted to improving the strength of pellets by increasing the efficiency of charge mixers [27].

The dosage of the polymer part in the BPC for each of the experiments did not exceed 0.5% of the bentonite mass. In each BPC, the polymer part is a single organic substance and not a mixture of them. This approach somewhat simplifies the description of the processes occurring during the interaction of BPC particles with water of iron ore concentrate and the surface of its particles and makes it possible to identify direct dependencies and the most promising polymer substances.

The green pellet technique used in all experimental laboratory studies is included in detail in [28] and is methodologically consistent with the method proposed by Eisel and Kawatra [29] and is still used at Michigan Technological University [30].

Before roasting, green pellets with a particle size of $-12.5 + 10$ mm were placed in trays and dried in an oven at a temperature of 105°C. Then, the dried pellets (sample weighing mp = 300 g) were placed in a basket made of nichrome wire. The pellets were fired in a laboratory high-temperature furnace VSP-220-1350 designed by NPVP "TOREX".

The furnace is a vertical cylinder, the interior of which is lined with refractory and heat-insulating material. The heating elements are silicon carbide rods (15 pcs.), installed vertically around the circumference of the furnace. An alundum glass is inserted into the furnace to protect the split heaters from destruction and average the temperature field in the working space. An oxidizing atmosphere is provided in the furnace (it is possible to supply both an inert gas to maintain a neutral atmosphere and oxygen or air through an opening in the bottom cover). A compressor is provided to pump air. The maximum temperature achieved in the furnace is 1300°C.

The hardware design of the furnace makes it possible to successfully carry out laboratory roasting under conditions of the process of thermal strengthening of pellets in existing roasting machines.

3. Research results and discussion

The following results from two experimental studies share the following characteristics:

- The produced pellets are non-fluxed.
- The polymer part of the bentopolymer composition is represented by high-viscosity polyanionic cellulose (PAC HV), a universal drilling additive with a high degree of polymerization and high molecular weight, which serves as a viscosity regulator to control fluid loss and inhibit clays in drilling fluids.

Results of experimental studies of the influence of BPC on the strength of roasted non-fluxed pellets of Stoilensky GOK:

During experimental studies of the effectiveness of the process of laboratory heat-strengthening roasting of iron ore pellets of the Stoilensky GOK using a bentopolymer composition, two samples of roasted pellets were obtained, produced in a laboratory with a bentonite binder of current production and with a bentopolymer composition (BPC) [28].

First of all, the strength characteristics of the resulting pellets were assessed, namely, the compressive strength of the roasted pellets, which was 392 and 350 kg/pellet for basic and bentopolymer pellets (with a 30% reduction in consumption), respectively. In the process of thermal strengthening of pellets, their necessary metallurgical characteristics are achieved both due to solid-phase sintering or ore phases and due to the formation of a melt during roasting at maximum temperature, which solidifies during cooling in the form of iron silicate glass and forms the structure of pellets of the required strength in the initial state and during the reduction process in blast furnace smelting. Moreover, if in the production of fluxed pellets, obtaining such an amount of melt (2–6%) does not cause difficulties, then in the case of non-fluxed pellets, obtaining such an amount of melt can be problematic [31].

The results of the study showed that a reduction in the consumption of the binder (represented by low-melting silicates) by 30% had a critical impact on the strength of the roasted pellets.

Thus, it has been established that the optimal dosage of BPC, which ensures an improvement in all qualitative and technological indicators of raw pelletization, does not provide an increase in the efficiency of heat-strengthening roasting for non-fluxed pellets from regrinded concentrate. However, the use of BPC had a positive effect on the chemical composition of the finished pellets, the mass fraction of Fe_{total} increased by 0.2%, and in turn, SiO_2 decreased by 0.19% (**Table 2**).

Reducing bentonite consumption seems necessary to optimize the chemical composition of finished pellets, which is especially important for high-quality pellets produced from re-enriched concentrates.

The results of chemical analysis showed that a significant reduction in bentonite consumption (by 30%) led to a decrease in silicon dioxide content by 0.2%, while the permissible error of chemical analysis (GOST 23581.15), depending on the technique, can range from 0.1 to 0.2%, i.e., the silica reduction values are within the acceptable error range. This suggests that a reduction in bentonite dosage by 1–2 kg/t can be practically

Name of binder	**Mass component, %**			
	Fe_{total}	**FeO**	**SiO_2**	**CaO**
Basic bentonite binder	65.9	0.29	4.94	0.22
BPC-1	66.1	0.22	4.75	0.20

Table 2.
Chemical composition of pellets produced with different binders.

undetected in the results of chemical analysis. Taking this into account, reducing the consumption of bentonite in the pelletizing process without losing the quality characteristics of the pellets remains an important task both from the point of view of reducing the cost of finished pellets by reducing the volume of purchased binder and from the point of view of the rational development of scarce and exhaustible high-quality bentonite.

Results of experimental studies of roasting non-fluxed pellets from high-quality iron ore concentrate from Sokolovsko-Sarbaiskiy mining and processing production association (GPO) in a laboratory furnace:

In this study, the qualitative characteristics of pellets produced using BPC were compared with the characteristics of pellets produced not with the binder additive used at the Sokolovsko-Sarbaiskiy mining and processing production association (SSGPO) pelletizing plant (variegated clay contained in overburdened rocks) but with the best and most promising bentonite clay of the commonwealth of independent states (CIS) known today—Taganskoe field, Kazakhstan. The unique rheological properties of bentonite clays from this deposit are due to the high content of the main mineral. According to X-ray diffraction and thermal analyses, the content of montmorillonite in the bentonite horizons of this deposit reaches 90–98% [32].

Thus, the effect of using BPC in the process of pelletizing and roasting was compared with the effect of using a bentonite binder, which was close to the reference one in its rheological parameters.

For both the bentopolymer composition and the bentonite binder, based on the results of green pelletization tests, the optimal consumption was determined to be 0.4% for each binder (**Table 3**). A further increase in binder consumption is impractical since, at the established consumption, the strength properties of the green and dried pellets met the technical requirements.

Thanks to the use of high-quality binders, iron ore pellets with high compressive strength and low abrasion resistance (B-0.5) were obtained (**Table 4**).

When using BPC, the porosity of the pellets increased by 1.1%, which led to an increase in the recoverability index according to ISO 4695 dR/dT, %/min, amounting to 0.70 for pellets produced with BPC and 0.66 for pellets produced with bentonite Taganskoye deposits. High rheological properties of binders make it possible to obtain high-quality green, dry, and roasted pellets. Thanks to the use of a bentopolymer composition, it is possible to optimize the chemical composition of the pellets with optimal binder consumption. The results of this study demonstrated that when using a bentopolymer composition and high-quality bentonite while reducing binder consumption, it is possible to obtain comparable quality characteristics of pellets. Thus, replacing a bentonite binder with a higher quality one or one with a bentopolymer composition should be considered as an equal option for implementing the idea of reducing binder consumption.

Results of pilot tests on the use of three different BPC compositions during pelletizing and roasting of DR pellets from re-enriched concentrate.

Binder	Compressive strength of green pellets, kg/pellet	Compressive strength of dry pellets, kg/pellet	Drop strength, n times H = 500 mm
BPC consumption 0.4%	1.29	5.71	2.5
Bentonite (Taganskoe) consumption 0.4%	1.27	6.31	2.6

Table 3.
Strength of green and dry pellets.

Binder	Binder consumption, %	Strength characteristics of roasted pellets			Porosity, %	Chemical composition on roasted pellets, %							
		On hit (B + 5), %	Abrasion resistance (B-0.5) %	Compression, kg/pellet		Fe_{total}	FeO	SiO2	CaO	MgO	Al_2O_3	S	Other
Bentonite	0.4	98.0	21.7	314.0	21.7	66.44	0.66	2.7	0.51	0.71	0.99	0.0018	0.19
BPC	0.4	98.1	22.8	318.0	22.8	66.34	0.94	2.68	0.52	0.70	0.99	0.0022	0.2

Table 4.
Characteristics of finished pellets.

The level of negative impact is an important resource aspect of the functioning of metallurgy. Emissions from metallurgical production constitute a significant share of gross national emissions from stationary sources and manufacturing in the Russian Federation.

Parameters such as the quality of iron ore raw materials, the level of environmental pollution by emissions of carbon-containing gases, and the energy efficiency of the pyrometallurgical process are interrelated. Improving the quality of the agglomerated product (primarily its strength) ensures a reduction in the recirculation load during processing. It is accompanied by a reduction in energy costs for production and a decrease in dust emissions [26].

According to some estimates [33], the market demand for steel produced using an alternative reducing agent (carbon-free method) in DR pellets by 2030 will be 15 times higher than the current production volume (2.25 million tons versus 147 thousand tons in 2019). For example, there is evidence in the literature that the use of hydrogen gas as a reducing agent for the direct reduction of iron ore materials will make it possible to obtain a carbon-free iron product with the prospect of reducing emissions of harmful gases, regardless of the type of reduction or smelting reactor used [34]. In addition to minimizing the negative impact on the environment, including risks associated with the global greenhouse effect problem, the interest of metallurgists in the processes of direct reduction of iron is caused by the limited and rapid depletion of coking coal reserves [35].

This study was conducted on high-quality concentrate intended for the production of DR pellets. Compared to ordinary pellets, DR pellets have a higher iron content and a low level of harmful impurities – silicon and sulfur. The idea of using BPC when working with super-concentrates is especially relevant because it is primarily aimed at reducing bentonite consumption. Since the purpose of iron ore processing is to remove silicate minerals, adding silicate as a binder is counterproductive.

The formulation of the studied binders differed both in the type of polymer additive used and slightly in the dosage of soda ash Na_2CO_3 into the original clay until a sufficient level of rheological properties of activated bentonite was achieved (**Table 5**). The dosage of the polymer additive is the same in each of the three binders. The basis of each of the studied binders is bentonite clay from the central part of the fifth layer of the 10th Khutor deposit (Republic of Khakassia).

For experimental samples of pellets, the metallurgical properties of direct reduction furnaces were determined (ISO 11257 and ISO 11258).

As can be seen from **Table 6**, with the use of BPC-7, it was possible to achieve an increase in all parameters recorded according to ISO 11257 and ISO 11258. It is

Name of binder	Swelling index, ml/2 g	Effective viscosity, mPa s	Swelling, times	Polymer additive type
Bentonite binder	30.7	40	15.2	None
BPC-4	30.04	73.5	14.4	Xanthan gum
BPC-6	34.4	99.0	14.9	High-viscosity polyanionic cellulose A
BPC-7	31.6	74.0	14.2	High-viscosity polyanionic cellulose B

Table 5.
Qualitative and rheological parameters of the studied binders.

Binder used	ISO 11257	ISO 11258			
	Determination of reduction-grinding index at low temperature and metallization index	**Determination of the reducibility index, final degree of reduction and degree of metallization, %**			
Defined parameter	RDIDR (<3.15MM)	M	dR/dt (40)	MR	R (90)
With base bentonite binder	1.03	92.55	1.25	83.1	8.2
With BPC-4	0.84	92.86	1.15	80.8	86.6
With BPC-6	1.76	91.43	1.39	88.7	92.1
With BPC-7	0.71	93.2	1.24	84.4	89.1

Table 6.
Results of the analysis of metallurgical characteristics of pellets for direct reduction.

interesting to compare the two extreme compositions because the polymer additive in BPC-6 and BPC-6 is represented by highly viscous polyanionic cellulose but from different manufacturers.

Of the negative consequences of using BPC-6:

- The reduction-grinding index at low temperature increases significantly from 1.03 to 1.76.
- The metallization rate also decreased from 92.55 to 91.43%.

At the same time, the reducibility index, the final degree of reduction, and the degree of metallization significantly increased relative to the basic values. However, during reduction, the strength of iron ore pellets decreases significantly. A sharp softening on the pellets during reduction can lead to their destruction in the furnace. The main factor influencing the behavior of pellets during reduction is the structure, which determines the rate of reduction of the pellets. The higher the specific surface area and the average pore size, the more likely the reduction process is to occur throughout the entire volume of the pellet, and the higher the reduction rate, the lower the strength and the higher the destructibility of the pellets.

Thus, it has been established that by using BPC as part of the charge for pelletizing, it is possible to influence the metallurgical properties of pellets for direct reduction iron (DRI), but even such parameters as the recovery rate have their optimal limits, since, on the one hand, increasing the rate of recovery of raw materials will increase the productivity of the DRI unit, but on the other hand, an excessive reduction rate leads to destruction of the pellets during reduction and subsequent problems in the operation of the furnace.

4. Conclusions and outlook

The mechanical and metallurgical properties of pellets are determined by many factors: the mineral composition of the feedstock, the amount and composition of fluxes, the structure of the pellets, the heat treatment regime, the size of the pellets, uneven firing across the layer and the nature of porosity. Controlling these factors at a pelletizing factory is possible if technical solutions are used aimed at modifying the macrostructure and preventing its defects.

The use of bentopolymer compositions as part of the charge is a promising direction for the production of iron ore pellets in order to improve the technical and economic performance of roasting machines.

An increase in the rheological parameters of the binders at all stages of the research made it possible to obtain high-quality green, dry, and roasted pellets. Thanks to the use of a bentopolymer composition, it is possible to optimize the chemical composition of the pellets with optimal binder consumption.

As a result of laboratory and semi-industrial research, it was revealed that it is possible to influence and quality characteristics of pellets using a bentopolymer binder. It should be especially noted that the dosage of the polymer substance to bentonite, which significantly increased the quality indicators of the pellets, was negligibly small.

The main scientific and technical achievements in the palletization of iron ore raw materials in recent years are somehow related to the optimization of the composition of the charge. The results of the studies also demonstrated that there is indeed a reserve for increasing the technological parameters of the pelletizing process. With this in mind, we expect continued academic and industrial research around the world to overcome the bentonite binder problem in a variety of ways.

The combination of properties of bentonite has ensured its practical irreplaceability in the pelletizing process over the past 60 years. The development of the science of clays and nanocomposites, which has been rapidly developing in recent decades, will inevitably lead to new discoveries, which will most likely ultimately lead to a revision of traditional approaches to the use of such a unique material as bentonite.

Author details

Daniil Vetyugov[1,2*] and Tamara Matveeva[1]

1 ICEMR RAS, Moscow, Russia

2 Bentonite Khakassia LLC, Chernogorsk, Russia

*Address all correspondence to: vetug7@gmail.com

References

[1] Hao L, Bing X, Yue-lin Q. Effect of bentonite on the pelleting properties of iron concentrate. Journal of Chemistry. 2017;**6**:1-6

[2] Mal'tseva VE, Seleznev VS, Chukin M. Pelleting and Preparation of Materials: Technology and Control Systems. M: Metallurgizdat; 2023

[3] Doroshenko MV, Bashlykova TV. Technological properties of minerals. In: Handbook for Technologists. M: Teploenergetik; 2007

[4] Guven N. Bentonites - clays for molecular engineering. Elements. 2009; **5**(2):89-92

[5] Bajestani MS, Nasir O, Oh WT. Properties of bentonite-based sealing materials during hydration. Minerals. 2023;**13**:1412

[6] Thakkar KS, Syed M, Yadav O, Krishan DS, Shukla MS. Application of bentonite clay as a binding material. Design Engineering (Toronto). 2021;**12**: 9111-9131

[7] Parathodiel H, Mousa E, Ahmed H, Elsadek M, Forsberg K, Andersson C. Developing iron ore pellets using novel binders for H2-based direct reduction. Sustainability. 2023;**15**: 11415

[8] Liu S, Zhang Y, Su Z, Jiang T. Improving the properties of magnetite green pellets with a novel organic composite binder. Materials. 2022;**15**: 6999

[9] De Moraes SL, Ribeiro TR. Brazilian iron ore and production of pellets. Mineral Processing and Extractive Metallurgy Review. 2019; **40**(1):16-23

[10] Osipov VI, Sokolov VN. Clays and their Properties. Compostion, Structure and Formation of Properties. Moscow: Geos; 2013

[11] Vetjugov AV, Voevodin LI, Mal'tseva VE. Bentonite Clay Activation Method. RU Patent 2199504 C1. 2003

[12] Forsmo SE, Apelqvist AJ, Björkman BT, Samskog P-O. Binding mechanisms in wet iron ore green pellets with a bentonite binder. Powder Technology. 2006;**169**(3): 147-158

[13] De Moraes SL, De Lima JB, Neto JF, Fredericci C, Fredericci SEM. Binding mechanism in green iron ore pellets with an organic binder. Mineral Processing and Extractive Metallurgy Review. 2020; **41**(4):247-257

[14] De Moraes SL, De Lima JB, Ribeiro TR. Iron ore pelletizing process: An overview. In: Iron Ores and Iron Oxide Material. London, UK: INTECH; 2018

[15] Claremboux V. Understanding strength of dried iron ore pellets [open access dissertation]. Michigan: Michigan Technological University; 2022

[16] Ngara T, Mavengere S, Bright S, Mapabma L. Investigating the effectiveness of organic binders as an alternative to bentonite in the pelletization of low grade iron ore. Physicochemical Problems of Mineral Processing. 2023;**59**(6):176094

[17] Zhuchkov VI, Zayakin OV, Akberdin AA. Prospects for using boron in metallurgy. Report 1. Izvestiya. Ferrous Metallurgy. 2021; **64**(7):471-476

[18] Halt JA, Kawatra SK. Iron ore pellet dustiness part II: Effect of firing route and abrasion resistance on fines and dust generation. Mineral Processing and Extractive Metallurgy Review. 2015; **36**(5):340-347

[19] Kawatra SK, Claremboux V. Iron ore pelletization: Part II. Inorganic binders. Mineral Processing and Extractive Metallurgy Review. 2021;**43**(7):813-832

[20] Pokolenko AY, Bragin VV, Gruzdev AI, Sharkovskiy DO, Puzakov PV, Izmagilov RI, et al. Analysis of the results and prospects for further comprehensive modernization of the pelletizing plant of JSC "MIKHAILOVSKY GOK named after a. V. Varichev" with roasting machines of the OKM-520 type. Steel. 2020;**11**:6-9

[21] Tsymbal VP, Kozhemyachenko VI, Rybenko IA, Mochalov SP, Padalko AG, Kalashnikov SN, et al. IV all-Russian scientific and practical conference with international participation. In: Prerequisites and Principles for Creating a Self-Organizing Jet-Emulsion Metallurgical Reactor. Novokuznetsk: Publishing Center of the Siberian State University; 2016

[22] Belousov PE, Krupskaya VV, Zakusin SV, Zhigarev VV. Bentonite clays from 10th khutor deposit: Features of genesis, composition and adsorption properties. RUDN Journal of Engineering Research. 2017;**18**(1):135-143

[23] GOST 21283-93. Bentonite Clay for Fine and Building Ceramics: Interstate Standard. Minsk: Publishing and Printing Complex of the Russian Gosstandard "Publishing House of Standards"; 1995. p. 8

[24] Kaufhold S, Dohrmann R, Ufer K, Meyer FM. Comparison of methods for the quantification of montmorillonite in bentonites. Applied Clay Science. 2002; **22**:145-151

[25] Belousov PE, Pokidko BV, Zakusin SV, Krupskaya VV. Quantitative methods for determining the content of montmorillonite in bentonite clays. Georesources. 2020;**22**(3):38-47

[26] Chizhikova VM, Bizhanov AM. Green Technologies in Modern Agglomeration. Vologda: Infra-Engineering; 2023

[27] Jensen JR, Murr DL. Roll Mixing Improves Binding of Cook Clay in Taconite Agglomeration. Society for Mining, Metallurgy & Exploration; 2006

[28] Vetyugov DA, Matveeva TN. Experimental study of the pelletizing process of reground iron ore concentrate from the Stoilensky GOK using new compositions of bentonire-polymer binders. Ferrous Metals. 2023;**3**:4-10

[29] Eisele TC, Kawatra SK. A review of binders in iron ore pelletization. Mineral Processing and Extractive Metallurgy Review. 2003;**24**:1-90

[30] Kawatra SK, Claremboux V. Iron ore pelletization: Part I. Fundamentals. Mineral Processing and Extractive Metallurgy Review. 2021;**43**(4):529-544

[31] Zhuravlev FM, Malysheva TY. Pellets from Concentrates of Ferruginous Quartzites. M: Metallurgy; 1991

[32] Sapargaliev EM, Kravchenko MM. Features of the genesis of the Taganskoye bentonite deposit in the Zaisan depression. Vestnik RUDN. Series: Engineering Research. 2007;**3**:40-46

[33] Teck Resources Limited. Steelmaking Coal Resilience. Presentation. 2023. Available from: teck.

com/media/Steelmaking-Coal-Resilience.pdf

[34] Htet T, Tan Z, Spooner S, Degirmenci V, Meijer K, Li Z. Gasification and physical-chemical characteristics of carbonaceous materials in relation to Hisarna ironmaking process. Fuel. 2021;**289**:119890

[35] Ekhebume OJ, Olufemi AO, Olugbemiga OO, Ofi O. Recent Trends in the Technologies of the Direct Reduction and Smelting Process of Iron Ore/Iron Oxide in the Extraction of Iron and Steelmaking. London, UK: IntechOpen; 2023